Quantitative Millimetre
Wavelength Spectrometry

RSC Analytical Spectroscopy Monographs

Series Editor: Neil W. Barnett, *Deakin University, Victoria, Australia*

Advisory Panel: F. Adams, *Universitaire Instelling Antwerp, Wirijk, Belgium*; M.J. Adams, *RMIT University, Melbourne, Australia*; R.F. Browner, *Georgia Institute of Technology, Atlanta, Georgia, USA*; J.M. Chalmers, *VS Consulting, Stokesley, UK*; B. Chase, *DuPont Central Research, Wilmington, Delaware, USA*; M.S. Cresser, *University of York, UK*; J. Monaghan, *University of Edinburgh, UK*; A. Sanz Medel, *Universidad de Oviedo, Spain*; R.D. Snook, *UMIST, UK*

The series aims to provide a tutorial approach to the use of spectrometric and spectroscopic measurement techniques in analytical science, providing guidance and advice to individuals on a day-to-day basis during the course of their work with the emphasis on important practical aspects of the subject.

Recent titles:

Industrial Analysis with Vibrational Spectroscopy, by John M. Chalmers, *ICI Research & Technology, Wilton, UK*; Geoffrey Dent, *Zeneca Specialities, Blackley, UK*

Ionization Methods in Organic Mass Spectrometry, by Alison E. Ashcroft, *formerly Micromass UK Ltd, Altrincham, UK; now University of Leeds, UK*

How to obtain future titles on publication

A standing order plan is available for this series. A standing order will bring delivery of each new volume immediately on publication. For further information, please write to:

Sales and Customer Care
Royal Society of Chemistry
Thomas Graham House
Science Park
Milton Road
Cambridge
CB4 0WF
UK

Telephone: +44(0) 1223 432360
E-mail: sales@rsc.org

RSC
ANALYTICAL
SPECTROSCOPY
MONOGRAPHS

Quantitative Millimetre Wavelength Spectrometry

John F. Alder
*Department of Instrumentation and Analytical Science, UMIST,
Manchester, UK*

John G. Baker
*Department of Physics and Astronomy, University of Manchester,
UK*

ROYAL SOCIETY OF CHEMISTRY

Chemistry Library

ISBN 0-85404-575-9

Published by The Royal Society of Chemistry,
Thomas Graham House, Science Park, Milton Road, Cambridge CB4 0WF, UK

Registered Charity Number 207890

For further information see our web site at www.rsc.org

Typeset by Keytec typesetting Ltd
Printed by MPG Books Ltd, Bodmin, Cornwall, UK

Preface

In this monograph the authors' aim is to demonstrate the current status and future potential of millimetre wavelength (MMW) spectrometry for quantitative analysis of gaseous mixtures. Spectroscopic theory is outlined in sufficient detail to form the basis of a model for the quantitative interpretation of the spectroscopic measurements. Details of the principal parts of the spectrometer are revealed and explained, permitting the analytical spectroscopist to specify and build a spectrometer from commercially available components. Quantitative models are developed for off-line signal processing and filtering to optimise the analytical performance.

We have concentrated on design concepts that would be acceptable to the modern community of analytical scientists. Compact, low-cost, automatic, robust instruments have been the goal. This has naturally excluded a number of devices and techniques that nonetheless would offer useful application in the research laboratory. Our focus has been on the high Quality Factor (Q) Fabry–Perot cavity spectrometers with frequency modulation (FM) of the spectral source and phase coherent detection of the absorption signals. These compact devices, exhibiting long absorption pathlength, are well suited to quantitative analytical work in process monitoring and control applications when part of an automatic computer controlled instrument.

Solid-state MMW sources with both cryogenically cooled and room temperature detectors are discussed, along with their application to conventional and dielectric perturbation based methods of quantification. Post-detection signal processing, smoothing, filtering and spectral profile fitting are described, which extract the full value from data obtained by this sensitive analytical technique.

The prospect of atmospheric pressure MMW spectrometry has attracted the authors throughout their studies. This monograph considers the potential application of higher pressure spectroscopic measurements at millimetre wavelengths and looks to the future application of those methods in millimetre wave-over-fibre technology.

The authors are grateful to the EPSRC in particular for funding much of their research over several decades, and to all the other agencies and companies for their support. Many thanks are due also to colleagues and students for their teaching and assistance, particularly to Gunnar Thirup. JGB would like to express his debt to the late Walter Gordy for introducing him to the power and the excitement of millimetre wavelength spectroscopy, and to Wenlie Liang, Senior Researcher at NPL, for the initial design and development of a cavity spectrometer. The efforts of Nick Bowring together with Jim Allen and his electronics workshop technicians were central to its successful implementation.

Many of the spectra shown in this book were obtained by Nacer-Ddine Rezgui, who played a very important role in our research into the scanning cavity spectrometer. This challenge has now been taken up enthusiastically by Ashley Wilks who assisted us also with proof reading; we appreciate them both.

Special thanks are due to Maureen Williams and Geraldine Smith for all their help over many years.

Contents

Glossary of Terms and Symbols

A.U.	arbitrary units
$A_{m \leftarrow n}$	Einstein coefficient of spontaneous emission from state n to m
$\arccos(x)$	the smallest angle whose cosine is x
α	sample absorption coefficient (m^{-1})
$\alpha_{\text{d,max}}$	absorption at peak of Doppler broadened line
$\alpha_{\text{max}}, \alpha_0$	absorption at central peak of Lorentzian broadened line
α_v	absorption at frequency v of Lorentzian broadened line
B	molecular rotation constant (MHz)
$B_{m \rightarrow n}$	Einstein coefficient of stimulated absorption from state m to n
BWO	backward wave oscillator
c	speed of light ($3 \times 10^8 \text{ m s}^{-1}$)
cgs	centimetre-gram-second system units
cm^{-1}	unit of frequency ($1 \text{ cm}^{-1} = 30$ GHz)
Dalton	atomic mass unit (C = 12)
dB	logarithmic power ratio in decibels (10 dB = 10× ratio)
$dF(v)/dv$	first derivative of spectral profile $F(v)$
$d^2 F(v)/dv^2$	second derivative of spectral profile $F(v)$
$\delta^2 F(v)$	second difference of function $F(v)$
$\Delta N_{m \rightarrow n}$	population difference between states m and n
ΔP	MMW power absorbed in traversing sample
Δv	spectral line half width parameter (MHz Pa^{-1})
Δv_{D}	Doppler broadened spectral line half width parameter
ESR	electron spin (paramagnetic) resonance
ε_0	permittivity of free space
$f_{m.r,v}$	fraction of molecules in state m, r, v
FASSST	fast scan sub-MMW technique
FT	Fourier transform procedure
FWHM	full width of spectral line (or cavity) at half maximum intensity
G	arbitrary multiplying constant
GaAs	gallium arsenide semiconductor material

GHz	10^9 cycles s^{-1}
γ, γ_{max}	sample absorption coefficient (cm^{-1})
h	Planck constant
HWHM	half width of spectral line (or cavity) at half maximum intensity
Hz	cycles s^{-1}
i	isotopic ratio in sample (p. 90)
i	current (Chapter 2)
I	moment of inertia (p. 17)
I	intensity (Chapter 1)
I	current (Chapter 3)
IF	intermediate beat frequency resulting from mixing two radio-frequency or MMW signals
j	imaginary number operator
J	rotational state quantum number
k	Boltzmann constant (throughout)
k	coefficient of coupling between cavity and waveguide (Chapter 2)
kHz	10^3 cycles s^{-1}
K	rotational quantum number in a symmetric rotor
L	litre
L	Avogadro's number (p. 10)
L	length (throughout)
L_{equiv}	equivalent transit path of radiation through sample
$\mathbf{L_0}$	inductive reactance of waveguide or transmission line
$\mathbf{L_c}$	inductive reactance of cavity
LO	local oscillator in MMW or RF frequency mixing system
LPG	liquefied petroleum gas
LTE	local thermodynamic equilibrium
λ	wavelength of MMW radiation
m	mass of molecule
M	relative mass of molecule (Dalton) (Chapter 1)
M	rotational quantum number (Chapter 1)
M	mutual inductance between two resonant circuits (Chapter 2)
MHz	10^6 cycles s^{-1}
MMW	millimetre wavelength
ms	10^{-3} s
$m|\mu|n$ or μ_{mn}	dipole moment matrix element for transition m to n
μ	molecular electric dipole moment
μs	10^{-6} s
N	concentration of molecules (m^{-3})
N_m	population of state m (m^{-3})
ν	frequency
ν_0	centre frequency of spectral line

ω	angular frequency ($= 2\pi\nu$ radian s^{-1})
P, P_{in}	incident MMW power on cavity or sample
P_{cav}	MMW power dissipated in cavity at resonance
P_{det}	MMW power reaching detector
Pa	SI unit of pressure (1 bar $= 10^5$ Pa)
PSD	phase sensitive detector or detection
Perspex	poly(methyl methacrylate) polymer material
PTFE	poly(tetrafluoroethylene) polymer material
p	pressure
p_1	standard or reference pressure
Q, Q_{L}	loaded Quality Factor of cavity
Q_0	unloaded Quality Factor of transmission line
Q_{c}	unloaded Quality Factor of cavity
Q_{s}	Quality Factor of sample
R_{c}	resistive impedance of cavity at resonance
RG99/U	waveguide dimension designator, in this case 3.1 mm $\times$ 1.5 mm internal dimensions
ρ	MMW amplitude reflection coefficient
ρ^2	MMW power reflection coefficient
$\rho_{m\to n}$	rate of transition from state m to state n
$\rho(\nu)$	energy density of radiation per unit frequency interval
Si	silicon semiconducting material
SI	international units of measurement based on metre-kilogram-second
S_{d}	Doppler spectral profile
$S(\nu)$	frequency dependent spectral profile
$\displaystyle\sum_{n=-N}^{N} n$	summation of values of n from $-N$ to $+N$
T	absolute temperature (Kelvin)
T, T_0	MMW power transmission coefficient
TEM$_{mnq}$	MMW oscillation mode with quantum numbers m, n and q, in a cylindrical cavity
TWT	travelling wave tube
torr	unit of pressure (1 orr $= 133.3$ Pa)
t	MMW amplitude transmission coefficient
t^2	MMW power transmission coefficient
τ	decay time of MMW signal in cavity
V_{cav}	MMW voltage appearing at cavity
V_{det}	MMW voltage appearing at detector
V_{w}	working voltage
w	half width of spectral line
x	fractional abundance

YIG   yttrium iron garnet tuneable MMW source

Z complex impedance of an alternating current circuit
Z_0 characteristic impedance of a waveguide or transmission line

CHAPTER 1

Interaction of Millimetre Wavelength Electromagnetic Radiation with Gases

Millimetre wavelength (MMW) radiation forms the 30–300 GHz band of the electromagnetic spectrum and is used to study the mainly rotational spectra of gaseous molecules. Transitions can also be measured in the inversion spectrum of ammonia, between molecular rotamer configurations and between magnetic fine structure components of molecules possessing a magnetic dipole, *e.g.* O_2 and NO. The absorption spectra normally studied in small molecules (<200 Dalton) become more intense at higher frequencies, as illustrated in Figure 1.1 for sulfur dioxide, so for quantitative work it is usually advantageous to work in the MMW band.

The rotational levels lie at low energies and are all populated at ambient temperatures. This gives rise to abundant spectra of narrow lines, width <1 MHz at Pascal (Pa) pressures, that are spread over this wide spectral region. Spectral interference between species in real mixtures is unusual and regions can be chosen for measurements where the target species are represented without spectral overlap. This is quite different from other spectrometric methods in the ultraviolet–visible to mid-infrared region where the analyst is usually constrained to work at one of a few frequencies. Separation prior to analysis for gas mixtures is not required apart perhaps from gross scrubbing of suspended solids or liquids, making rotational spectrometry unique amongst quantitative analytical methods.

Rotational spectroscopy is essentially a low-pressure technique if one is to exploit its remarkable selectivity, although quantitative measurements can be made at pressures up to atmospheric. Not all gases are rotationally active; with the exception of dioxygen all the homonuclear diatomic molecules and of course monatomic gases are inactive. Molecules with a high degree of symmetry, notably methane, ethane, ethene, benzene and carbon dioxide, are likewise inactive or have weak spectra, *e.g.* propane and butane. Oxygen and water, possibly the two most-determined molecular species in the modern world, are rotationally active and can be measured readily in those matrix gases.

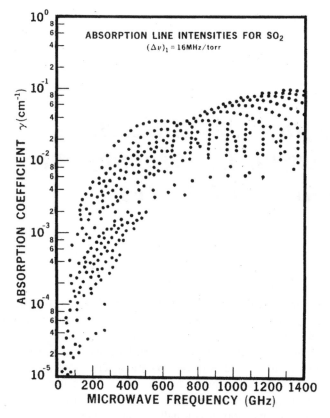

Figure 1.1   *Absorption coefficients for sulfur dioxide computed by Kolbe and Leskovar.[7]*
(Reprinted from Kolbe and Leskovar[7] with permission from Elsevier Science)

Modern teaching in analytical spectrometry deals only rarely with quantitative rotational spectrometry in any depth and this lack of attention gives rise to some misunderstandings about the technique. Much of that derives from the unusual and sometimes seemingly mysterious combination of optical and electronic phenomena that characterise this spectral region. In reality of course, MMW spectrometry is quite simple and has been made much easier in recent years with the advent of solid state electronic instrumentation and devices of very high quality.

The theory too is fundamentally simple and has been very well developed although less emphasis has been placed on the quantitative aspects of the subject than was perhaps desirable, with the notable exception of the book devoted to this topic by Varma and Hrubesh.[1]

The origin and details of molecular rotational spectra are explicitly laid out in the comprehensive texts by Townes and Schawlow,[2] and Gordy and Cook,[3] the latter two being the sources for much of the theory outlined below. The present authors have focused only upon those aspects that are directly relevant to quantitative MMW cavity spectrometry, and the reader is referred to the original texts for a more comprehensive treatment. Readers will soon notice that the original texts, and indeed the majority of all published MMW spectrometry papers and texts to this day, use cgs rather than SI units. Consequently, some of the equations differ from those in other texts and translation of units will be necessary to compare them with other work.

1 Basic Spectroscopic Theory

The purpose of this section is to give readers access to the most important relationships between the spectroscopy and the quantitative measurement of molecular species in mixtures. The theory aims to form a relationship between the composition of a gas mixture and the peak intensity α_{max} of a spectral transition. The value of α_{max} is influenced by pressure broadening, temperature effects and power saturation, and takes on different forms with respect to the size and structure of the molecule. These factors all influence the spectral line shape that is directly related to the fractional abundance of that absorbing species in the sample being measured. The theory also permits at least semi-quantitative design of experimental parameters such as optimum working pressure, measurement frequencies for optimum sensitivity and location with concomitant gas absorption lines. It will help understanding of some design requirements and limitations of the technique. Most of all it is intended to remove some of the mystique from the subject. The initial goal is therefore to equate the maximum absorption coefficient α_{max} to the species' spectroscopic parameters and fractional abundance. It will be approached from first principles, with an eye focused always on the quantitative aspects of the theory.

The main difference between optical and MMW spectrometry lies in the fact that energies involved in the rotational transitions are $\sim$100–1000 times less than those involved in the usually more familiar vibrational and electronic transitions of molecular and atomic species.

The Boltzmann equation exponent $h\nu/kT$ at ambient temperatures is approximately $4.8 \times 10^{-3}\ \sigma/\mathrm{cm}^{-1}$ or $1.6 \times 10^{-4}\ \nu/\mathrm{GHz}$. Thus for rotational energy levels lying typically in the tens-hundreds cm^{-1} region[*] above the ground state the exponential term is not much less than unity, indicating that most levels will hold a significant fraction of the overall population of species. This has mostly positive consequences through rich well-distributed spectra and a choice of frequency region for observation of particular gas mixtures.

If a molecule possessing two rotational energy levels E described by the quantum numbers m for the lower level and n for the upper level, is exposed to radiation of frequency ν where

$$h\nu = E_n - E_m \tag{1.1}$$

---

[*]Although the SI unit for energy is J, energy level values are almost universally reported in *wavenumbers*, σ/cm^{-1}, in rotational spectroscopic literature; 1 cm^{-1} = 1.9864×10^{-23} J.

and h is Planck's constant, a transition will take place with probability

$$p_{m \to n} = \rho(\nu) B_{m \to n} \qquad (1.2)$$

where $p_{m \to n}/s^{-1}$ is the rate of change of the probability that the molecule will be found in the upper state; $\rho(\nu)$ is the energy density of the radiation per unit frequency interval$/J\,m^{-3}\,s$ and $B_{m \to n}/J^{-1}\,m^3\,s^{-2}$ is the Einstein coefficient of absorption for that transition.

In the excited state n the interaction of the radiation field with the molecule induces emission, and the molecule relaxes to the lower state m. The induced emission is indistinguishable from the field that caused it. Furthermore, it is of no use analytically unless the excitation source radiation is interrupted before the upper state population has had time to relax. That becomes the case in Pulsed Fourier Transform microwave spectroscopy where the excitation source is pulse modulated and the induced radiation is emitted against a very low microwave background.

The Einstein coefficients for induced emission and absorption are identical B_{mn} and can be expressed as

$$B_{mn} = (2\pi^2/3\varepsilon_0 h^2)[|(m|\mu_x|n)|^2 + |(m|\mu_y|n)|^2 + |(m|\mu_z|n)|^2] \qquad (1.3)$$

The terms in $\mu_{mn}/C\,m$ are the matrix elements of the molecule's dipole moment projected onto axes x, y, z fixed in space rather than onto the molecule itself; $\varepsilon_0/J^{-1}\,C^2\,m^{-1}$ is the permittivity of free space.

In MMW spectrometry the radiation is commonly plane polarised as it passes through the absorbing gas. This effect can give rise to alterations in the intensity of observed spectral transitions depending upon the extent of that polarisation and its relationship with other electrical and magnetic fields that may be introduced. Indeed, the effect is exploited to great advantage in Zeeman *magnetic field* modulation for the determination of radicals, and for Stark *electric field* modulation of the absorption line frequency; see Chapter 5. The Zeeman effect is noticeable in oxygen, where the energy levels are split in the Earth's magnetic field, and the intensity of the split transitions can be altered depending upon the orientation of the cavity in the laboratory, and the polarisation of the MMW radiation. This analytically unwelcome splitting can be easily overcome by magnetic shielding of the cavity (Figure 1.2).

Another route to depopulation of the upper level is spontaneous emission, which is described by an analogous spontaneous emission coefficient $A_{m \leftarrow n}/s^{-1}$. This mechanism is the basis of atomic emission spectrometry in the ultraviolet–visible spectrum particularly. In the MMW region with its much lower frequency transitions it is not a significant contributor, as will be demonstrated below. Modifying Equation 1.2 to bring in the populations N of the upper and lower states one can write for the absorption and emission processes

$$d/dt(\Delta N_{m \to n}) = N_m B_{mn} \rho(\nu) \qquad (1.4)$$

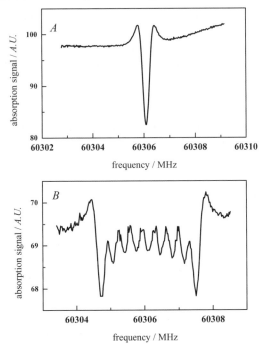

Figure 1.2 *Oxygen spectral line with A the cavity magnetically shielded and B the cavity unshielded, showing the line split by the Earth's magnetic field*
(Reprinted from J.F. Alder and J.G. Baker, *Anal. Chim. Acta*, 1998, **367**, 245–253, with permission from Elsevier Science)

and for the reverse process

$$d/dt(\Delta N_{m \leftarrow n}) = N_n[B_{mn}\rho(\nu) + A_{m \leftarrow n}] \tag{1.5}$$

If the two processes are in thermodynamic equilibrium these rates will be equal. The ratio of the populations of the two levels N_n/N_m will equate to the Boltzmann relationship and conform to Planck's radiation law.[4] From this can be derived the relationship

$$A_{m \leftarrow n}/B_{mn} = 8\pi h\nu^3/c^3 \ /J\,s\,m^{-3} \tag{1.6}$$

The strong frequency dependence of A/B shows why even though spontaneous emission is analytically useful in the ultraviolet–visible spectrum, it is not at MMW frequencies. At 300 nm the ratio is 6.17×10^{-13} J s m^{-3}, whereas at 300 GHz it is 1.67×10^{-23} J s m^{-3}, with typical lifetimes $1/A_{mn}$ of 10^{-8} s for ultraviolet visible, and 100 s for MMW emitter upper states.

It is normal therefore to work in the absorption mode to make quantitative measurements at MMW frequencies and the absorption coefficient α can be derived from a consideration of the power absorbed by a volume V of gas containing N molecules per unit volume. The number of molecules per second undergoing the transition $m \rightarrow n$ is the product of the total number of molecules and the probability of the transition taking place $p_{m \rightarrow n}$. As each photon has an energy $h\nu$ the total power absorbed from the field is, from Equation 1.2:

$$P_{m \rightarrow n} = VN_m\rho(\nu)B_{mn}h\nu \tag{1.7}$$

The radiation field also stimulates emission from the upper state n and the radiation returned to the field by this process is described by the analogous equation

$$P_{m \leftarrow n} = VN_n\rho(\nu)B_{mn}h\nu \tag{1.8}$$

and the nett change in power in the radiation field will be given by subtracting Equations 1.8–1.7:

$$\Delta P = V(N_n - N_m)\rho(\nu)B_{mn}h\nu \tag{1.9}$$

This equality underlines the important feature that the power absorbed is directly proportional to the difference in populations of the two levels involved in the transition $(N_n - N_m)$. That has some important implications for MMW spectrometry in the light of both the absolute values of the energy levels and the difference in energy between them, as is discussed below.

Under conditions of local thermodynamic equilibrium (LTE) thermal relaxation processes will be rapid and maintain the population of the lower level. If, however, one were able to depopulate the upper level n compared with its LTE population, the absorption signals would increase. The technique of double-resonance spectrometry exploits this by application of an intense MMW or laser field that excites a transition, *e.g.* $n \rightarrow q$, where q is a higher state (Figure 1.3). This depletes the population of n and permits greater absorption of a second MMW field, exciting the transitions $m \rightarrow n$ or $q \rightarrow r$.

Freezing the gas to a low temperature increases the population of state m with respect to n both in absolute and relative terms. This is of particular consequence as levels nearer to the ground state are measured. Significant sensitivity enhancement can be obtained by freezing the gas down to a few K rotational temperature as the absorption measurement is carried out in, for example, a supersonic jet.[5] There is a trade-off, however, between population increase of the lower states by depopulation of the higher states, because the intermediate states' populations too will increase as the rotational temperature falls. This is reached in OCS, for example, at $j = 5$ and $j = 6$ at 4 K.

The term $(N_n - N_m)$ can be expressed using the Boltzmann equation, assuming LTE:

$$N_n - N_m = N_m[1 - \exp(-h\nu/kT)] \tag{1.10}$$

Equation 1.9 becomes

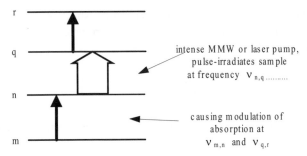

Figure 1.3 *Scheme showing the basis for double resonance spectrometry. The pump radiation could also excite higher vibrational states, thus depleting the lower vibrational state rotational levels, and increasing population of the higher state rotational levels. Both laser and MMW pumping would lead to modulation of the probe MMW absorption between the affected rotational levels. The chances of two different molecules having two coincident transition frequencies is very small, thus imparting high selectivity to this spectrometric approach*

$$\Delta P = VN_m[1 - \exp(-h\nu/kT)]\rho(\nu)B_{mn}h\nu \qquad (1.11)$$

Equation 1.11 can be made more practically useful by defining the absorption coefficient α as the fractional change in power per unit path-length Δl in the cell

$$\alpha = -(\Delta P/P)/\Delta l \ /\mathrm{m}^{-1} \qquad (1.12)$$

The volume and power density terms can be rationalised also: a volume element in the cell can be expressed as $A\Delta l$ where A is the cross-sectional area of the element. The energy density $\rho(\nu)$ can be expressed in terms of A and P by considering the wave-front from a monochromatic source propagating through the cell at velocity c:

$$\rho = P/cA \qquad (1.13)$$

The term N_m is more usefully expressed as the total concentration of molecules of that species N multiplied by the fraction in the lower state of the MMW transition f_m. Combining Equations 1.11, 1.12 and 1.13 yields

$$\alpha_{mn} = (Nf_m/c)[1 - \exp(-h\nu/kT)]B_{mn}h\nu \qquad (1.14)$$

The exponential term can be expanded because $h\nu < kT$ in the longer MMW region:

$$h\nu/kT = 0.048\nu(/\mathrm{GHz})/T(/\mathrm{K}) \qquad (1.15)$$

although at higher frequencies and low temperatures, this assumption should not be automatic: at 100 GHz and 10 K, $h\nu/kT = 0.48$.

Retaining the second term of the expansion for inspection shows that at low temperature and high frequency measurements, it can have a significant influence on the calculated absorption coefficient:

$$\alpha_{mn} = \frac{Nf_m(h\nu)^2}{ckT}\left(1 - \frac{1}{2}\frac{h\nu}{kT} + \ldots\right)B_{mn} \qquad (1.16)$$

but for most practical purposes it can be truncated at the first term of the expansion

$$\alpha_{mn} \approx \frac{Nf_m(h\nu)^2}{ckT}B_{mn} \qquad (1.17)$$

Equations 1.16 and 1.17 are based on the assumption that all the molecules undergoing the transition $m \rightarrow n$ do so at the same frequency ν. In reality they will have slightly different transition frequencies centred around the centre frequency ν_0 due predominately to collisional interactions between molecules. Doppler broadening also makes a small contribution giving a Gaussian shape to the line (Figure 1.4), but the overall result is a profile approximated by the Lorentz shape function $S(\nu)$:

$$S(\nu) = \Delta\nu/\pi[(\nu_0 - \nu)^2 + \Delta\nu^2] \qquad (1.18)$$

where ν is close to ν_0 and $\Delta\nu$ is the half-width of the spectral profile measured at the half-intensity points (HWHM).

B_{mn} in Equation 1.17 can be replaced in terms of the sum of the squares of the dipole moment matrix elements (Equation 1.3) and the equation modified by the Lorentz function, yielding the general form of the equation for α_ν the absorption coefficient at any point on the spectral profile:

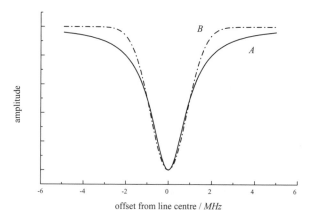

Figure 1.4   *Plot of the Lorentzian A, and Gaussian B, shape functions, both 2 MHz full-width at half-maximum amplitude*

$$\alpha_v = \frac{2\pi N f_m v^2}{3\varepsilon_0 ckT} \mu_{mn}^2 \frac{\Delta v}{(v_0 - v)^2 + \Delta v^2} \tag{1.19}$$

The integrated line absorption intensity I

$$I = \int_{-\infty}^{\infty} \alpha_v \, dv \tag{1.20}$$

is independent of line width. Integrated absorption intensity measurements should and largely do compensate for the varying broadening effects that may occur as the composition of the analyte mixture changes, and are the norm for quantitative MMW spectrometry.

When $v = v_0$ the centre frequency and maximum of the line profile, the peak absorption coefficient α_{max} is given by

$$\alpha_{max} = \frac{2\pi N f_m v^2}{3\varepsilon_0 ckT \Delta v} \mu_{mn}^2 \tag{1.21}$$

The expression above must be summed over all possible transitions between states m and n to obtain a total absorption coefficient. In rotational spectroscopy, every transition originating in state J consists of $2J + 1$ overlapping transitions whose magnetic quantum number M ranges from $-J$ to $+J$, but whose frequencies are identical. For plane polarised MMW interacting with all molecules, the relative strength of each M component works out as $1 - M^2/(J + 1)^2$ and, after summation, their nett contribution to the transition dipole strength becomes $(J + 1)\mu^2$, where μ is the molecular dipole moment.

In applying this formula to the determination of the peak absorption coefficient, it is customary to combine with it the rotational partition function contribution kT/hB giving the term $hB(J + 1)\mu^2/kT$, (ref. 3, p. 117). This equates to $hv\mu^2/2kT$ in linear and symmetric top molecules and replaces the μ_{mn} term in Equation 1.21, with f_m becoming f_J. This latter term takes into account the Boltzmann occupancy factor for that state: $f_J = \exp - [BJ(J + 1)/kT]$

$$\alpha_{max} = \frac{\pi h N f_J v^3 \mu^2}{3\varepsilon_0 c(kT)^2 \Delta v} \tag{1.22}$$

The inverse dependence of α_{max} on peak width Δv and its independence of pressure are clearly important and the proportionality to v^3 indicates the advantage of working at higher frequencies. Although there is a temperature dependence, this is unlikely to be significant under most near-ambient analytical measurement conditions.

2 Effect of Spectral Line Broadening on Absorption Intensity

Doppler Broadening

Doppler broadening is not an important factor in the lower MMW frequency region at ambient temperatures or below. At higher MMW frequencies its contribution to the overall linewidth could, however, become significant. Its effect can be readily demonstrated by recalling that the relative frequency shift in the spectral absorption due to the velocity of the molecule v with respect to the direction of the MMW radiation is:

$$v/c = (\nu - \nu_0)/\nu_0 \tag{1.23}$$

The velocity of the molecules is described by the Maxwell distribution (ref. 4, p. 26) and so the line shape function S_d can be considered as the unbroadened function S_0 modified by the shape of the Maxwell distribution:

$$S_d = S_0 \exp\left(-\frac{mc^2}{2kT}\left[\frac{\nu - \nu_0}{\nu_0}\right]^2\right) \tag{1.24}$$

where m is the mass of the molecule.

Equation 1.24 can be expressed in terms of the HWHM $\Delta\nu_d$ of the shape function S_d and its maximum amplitude $S_{d(max)}$. When $\nu = \nu_0 + \Delta\nu_d$ then the amplitude of the shape is $\frac{1}{2}S_{d(max)}$.

Noting that at $S_{d(max)}$, $\nu = \nu_0$ and taking the ratio $S_{d(max)}/\frac{1}{2}S_{d(max)}$, one obtains Equation 1.25:

$$2 = \exp\frac{mc^2\Delta\nu_d^2}{2kT\nu_0^2} \tag{1.25}$$

which can be rearranged to:

$$2\Delta\nu_d = \frac{2\nu_0}{c}\sqrt{\frac{2(\ln 2)LkT}{M}} \tag{1.26}$$

$2\Delta\nu_d$ is the full-width at half-maximum *FWHM*; L is Avagadro's number and M is the relative molecular mass of the molecule. Substituting values for the constants yields:

$$2\Delta\nu_d = 7.15 \times 10^{-7}\nu_0\sqrt{\frac{T}{M}} \tag{1.27}$$

For example, at 300 K, a gas of $M = 100$ Dalton and $\nu_0 = 173$ GHz would have a Doppler FWHM, $2\Delta\nu_d \sim 215$ kHz.

Equation 1.26 can be rearranged in terms of v_0 and substituted into Equation 1.24 to give a shape function in terms of the Doppler width of the line:

$$S_d = S_0 \exp\left[-\ln 2\left(\frac{v - v_0}{\Delta v_d}\right)^2\right] \tag{1.28}$$

This can be substituted into Equation 1.17 in the same way as was the Lorentz function, to give a value for the absorption coefficient of the Doppler broadened line profile.

It is appropriate at this point to express the term N, the number of molecules per unit volume, used in Equations 1.15–1.21, as a pressure. The term N/m^{-3} can be derived from the molar volume of a perfect gas and is, with p/Pa and T/K,

$$N = 7.24 \times 10^{22} \, p/T \tag{1.29}$$

Making these substitutions into Equation 1.19, and for clarity replacing the constant terms by a single constant G which is independent of both pressure and frequency, an expression is obtained for α_d which takes into account the Doppler contribution alone:

$$\alpha_d = Gv^2 p \exp\left[-\ln 2\left(\frac{v - v_0}{\Delta v_d}\right)^2\right] \tag{1.30}$$

The significance of Equation 1.30 is that at the centre frequency v_0 of the Doppler broadened line, the maximum absorption coefficient $\alpha_{d,max}$ is always independent of the Doppler width. As long as the pressure is below a critical value p_c, which corresponds to the onset of a significant contribution due to collisional broadening, then

$$\alpha_{d,max} = Gv^2 p_c \tag{1.31}$$

For pressures below p_c the peak absorption coefficent decreases linearly with pressure, but the Doppler linewidth does not (Equation 1.28), so there is no advantage in terms of spectral resolution in working below this critical pressure and a nett disadvantage in terms of sensitivity. Above p_c, $\alpha_{d,max}$ becomes constant and the true linewidth increases linearly with pressure.

This effect is only of real consequence at the higher MMW frequencies. Gordy and Cook (ref. 3, p. 47) calculate for example that for the 572.053 GHz NH_3 line at 300 K the Doppler width $2\Delta v_d = 1.7$ MHz, comparable with collisional broadening at ~ 13 Pa. In the 24 GHz region, the ammonia line Doppler widths are ~ 72 kHz, much less than the collisional broadening at typical working pressures.

The practically observed combined Doppler plus Lorentz profile is called the Voigt profile and can be computed numerically if required, to resolve the measured linewidth into its contributory components (ref. 6, p. 167).

Natural Linewidth and Pressure Broadening

The natural linewidth of a molecular spectral line in the MMW region is re-lated inversely to the spontaneous emission coefficient of the upper state A_{mn} (Equation 1.5) and consequently imparts a Lorentzian shape to the line profile (Equation 1.32). As the upper state can radiate to more than one lower state, the actual natural linewidth is related to the sum of the squared dipole moment matrix elements of the states involved.[3] In any case its contribution to the overall linewidth is negligible in comparison with the other broadening contributions: at 100 GHz it would be $\sim 10^{-5}$ Hz.

Collisions between molecules are the greatest cause of line broadening at the pressures normally employed for MMW spectrometry. In the Lorentz theory (ref. 2, p. 338) the lifetime of the rotational state involved in the transition is ended abruptly by collision with another molecule which stops the rotation. When the molecule starts to rotate again, its phase with respect to the other molecules is random. For an assembly of molecules this will give rise to an absorption line profile with a FWHM of $1/2\pi\tau$, where τ is the mean time between collisions. This is the linear sum of two terms, one for the upper and one for the lower state, having the shape of the Lorentz function (Figure 1.4) when $\Delta\nu \ll \nu_0$:

$$S(\nu, \nu_0) = \frac{1}{\pi} \left[\frac{\Delta\nu}{(\nu_0 - \nu)^2 + \Delta\nu^2} \right] \tag{1.32}$$

The origin of Equation 1.32 can be justified thus: the collision frequency is proportional to the pressure, as is the number density of absorbing molecules. Hence $\Delta\nu$ will be proportional to p. This proportionality is conventionally introduced into the equation by reference to a standard state (ref. 4, p. 819), where $\Delta\nu_1$ is the width at $p_1 = 1/760$ atm (133 Pa) and $T_1 = 300$ K. Using the relationship

$$\frac{T_1 \Delta\nu_1}{p_1} \approx \frac{T\Delta\nu}{p} \tag{1.33}$$

where the symbols not subscripted refer to the different temperature and pressure. The collision rate is proportional to the product of the mean relative speed of the molecules and their collision cross-section. The former quantity is proportional to $\sqrt{T}$. The collision cross-section temperature dependence is less well defined, being species dependent, but can be generalised to $\sqrt{T}$ also, thus giving the approximate relationship formalised in Equation 1.33.

Equation 1.19 can be expressed in terms of α_{max} (Equations 1.32 and 1.33) as

$$\alpha_\nu = (\alpha_{max} \nu^2 \Delta\nu / \nu_0^2) \left[\frac{p^2 \Delta\nu_1}{(\nu_0 - \nu)^2 + p^2 \Delta\nu_1^2} \right] \tag{1.34}$$

At the line peak, when the absorption coefficient is maximum, $\nu = \nu_0$ and the peak absorption coefficient is independent of pressure.

A more rigorous description of the pressure dependence of line shape can be obtained, that is useful at higher pressures. At the end of the collision that terminated the life of the excited state, the molecule was not in fact oriented at random, but was oriented in a direction at LTE with the MMW electric field existing immediately after each collision, in accordance with the Boltzmann distribution

$$\exp(-\boldsymbol{E}.\boldsymbol{\mu}/kT) \tag{1.35}$$

where $\boldsymbol{E}$ is the electric field strength existing at that instant and $\boldsymbol{\mu}$ is the dipole moment of the molecule, both being vector quantities (ref. 2, p. 339). This gives rise to the van Vleck–Weisskopf line shape that has the shape function

$$S(\nu, \nu_0) = \frac{\nu}{\pi\nu_0}\left[\frac{\Delta\nu}{(\nu_0 - \nu)^2 + \Delta\nu^2} + \frac{\Delta\nu}{(\nu_0 + \nu)^2 + \Delta\nu^2}\right] \tag{1.36}$$

Making the substitutions that gave Equation 1.34 now gives for the modified peak absorption coefficient α'_{max}

$$\alpha'_{max} = \alpha_{max}\left[1 + \frac{p^2\Delta\nu_1^2}{4\nu_0^2 + p^2\Delta\nu_1^2}\right] \tag{1.37}$$

Equation 1.37 shows that α'_{max} hardly changes from the Lorentz case, as the right-hand term in the bracket is dominated by the $1/\nu_0^2$ term, but the line becomes asymmetric, broadening towards the higher frequency as shown for water in Figure 1.5. This effect is of some consequence if analysis of gaseous mixtures at pressures around atmospheric is considered (Section 4.3 and Chapter 5).

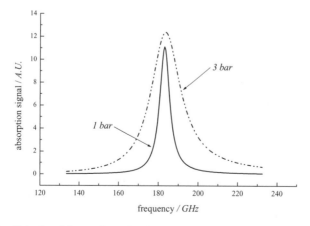

Figure 1.5 *Calculated Spectral Profile of the H$_2$O 183 GHz Line in Air at 1 bar, and at 3 bar. The asymmetry is evident, with broadening towards higher frequency*

Another aspect of collisional broadening is that due to collision with the cell walls. For a rectangular waveguide this can be approximated to

$$2\Delta\nu = \frac{1}{10}\left(\frac{1}{x} + \frac{1}{y} + \frac{1}{z}\right)\sqrt{\frac{T}{M}} \tag{1.38}$$

where x, y, z are the dimensions/m, M/Dalton and $\Delta\nu$/kHz (ref. 3, p. 52). It can be seen that for cavity spectrometers with absorption cell dimensions typically tenths of a metre, this effect will be usually well below 5 kHz and can be considered negligible.

Power Saturation and Broadening

When a gas in an absorption cell is first irradiated at a transition frequency there is a transient imbalance in the population of the two levels as molecules absorb the radiation and the upper level becomes over-populated. Thermal relaxation processes may not be sufficiently fast to repopulate the lower level. Subsequent power absorption will become less than expected from LTE considerations. This effect will be greater at the line centre where the absorption coefficient is greatest and can lead to complete saturation of the transition. The rate of absorption of radiation is then dictated solely by the rate of relaxation of molecules from the upper to the lower state. This is more noticeable at low pressures where the thermalisation processes are less rapid and the line widths narrower.

The result of this power broadening or saturation is to reduce the absorption in the line centre compared with the absorption in the wings of the line. This in turn leads to loss of analytical signal intensity and an apparent broadening of the absorption line profile. The resulting effect on the line shape function can be described by an equation due to Karplus and Schwinger, for low powers and incomplete saturation (ref. 3, p. 50):

$$S(\nu, \nu_0) = \frac{1}{\pi}\left[\frac{\Delta\nu}{(\nu - \nu_0)^2 + (\Delta\nu)^2 + B_{mn}P/c\pi^2}\right] \tag{1.39}$$

The term $P/\mathrm{J\,s^{-1}\,m^{-2}}$ is the *power density* influencing the molecule. It is identical to the intensity I and is related to the electric field strength $E/\mathrm{V\,m^{-1}}$:

$$I = \varepsilon_0 c E^2 \tag{1.40}$$

By setting $\nu = \nu_0$ in Equation 1.39 it can be seen that the effect of the power broadening is to reduce the value of S and hence α_0 at the line centre. The influence on α_0 is not great although the Q of the cavity does amplify the effect. Relatively little attention appears to have been given to this amplification in the literature, and so in the next section we derive expressions for the conditions under which it may occur.

Effect of the Cavity Q on Power Broadening

In this work we have focused on high-Q absorption cavities for spectrometric measurements, but the Karplus and Schwinger equation effectively refers to the gas being confined in a structure of unity-Q. The effect a high-Q resonant cavity has on the power density P can be best envisaged from a quasi-optical picture of the cavity behaviour. Power entering the cavity P_{in} through a coupling iris is reflected many times between the two mirrors of the cavity (Section 2.1). In order to traverse the length of the cavity to the exit coupling-iris, a particular wavefront has therefore spent an extended time in the cavity. During that period more energy has entered the cavity, which thus acts as a store of energy. The residence time is associated with a large number of reflections at each mirror but because the mirrors are not perfect some of the energy is lost, principally because of resistive heating of the mirror surfaces and by other mechanisms including absorption in the gas phase. The power in the wavefront thus decays naturally with a time constant τ related to that loss.

The *Quality Factor Q* is variously described, but most usefully in this context as

$$Q = \omega \text{ (energy stored/power lost by cavity)} \tag{1.41}$$

where $\omega/\text{rad s}^{-1}$ is the angular frequency of the MMW radiation.

The energy stored cannot increase *ad infinitum*, so ultimately the power input to the cavity must equal the sum of the power dissipated resistively and lost by other mechanisms: diffraction losses, coupling into the circuit *etc.* The power lost in the steady state is thus the input power P_{in}.

The electric field strength in the cavity E is not a singular quantity, it varies around the structure at resonance both spatially and temporally, but can be represented by a mean value $\bar{E}$, and is related to the energy stored by

$$\text{Energy stored} = \varepsilon_0 \bar{E}^2 V_{cav}/J \tag{1.42}$$

Rewriting the term ω, Q now becomes

$$Q = \frac{2\pi c \varepsilon_0 \bar{E}^2 V_{cav}}{\lambda P_{in}} \tag{1.43}$$

V_{cav} is the effective cavity volume that is occupied by the electric field. The electric field in a Fabry–Perot cavity operating in TM_{00n} mode is largely confined to a cylindrical element of radius $\sim(\lambda L/2\pi)^{\frac{1}{2}}$ and length L equal to the mirror spacing (Section 2.1). The term $c\varepsilon_0 \bar{E}^2$ equates to the power density term P in the Karplus and Schwinger equation. The power broadening contribution in that equation therefore becomes for a Fabry–Perot cavity

$$\frac{B_{mn} Q P_{in}}{c\pi^3 L^2} /\text{s}^{-2} \tag{1.44}$$

In applying Equation 1.44 to power saturation it needs to be remembered that B_{mn} is proportional to $1 - M^2/(J + 1)^2$ for each M component. This means that states of high M are both less intense and less readily saturated than those near $M = 0$. Although no simple formula can be applied to extract the resulting line profile, it is probably a good approximation to use the most intense component as a signal of the onset of power saturation.

To assess the importance of power broadening on a line of a molecule having $\mu_{mn} = 0.5$ D (1 D $= 3.34 \times 10^{-30}$ C m), consider a source generating $P_{in} = 100$ μW into a cavity length 0.5 m having $Q = 10^5$, working at 150 GHz and $\sim$13 Pa:

$$B_{mn} = \frac{2\pi^2}{3} \frac{|\mu_{mn}|^2}{\varepsilon_0 h^2} \tag{1.45}$$

where ε_0 is 8.85×10^{-12} J^{-1} C^2 m^{-1}, $h = 6.63 \times 10^{-34}$ J s, yielding a value for $B_{mn} = 4.3 \times 10^{17}$ J^{-1} m^3 s^{-2}.

The effect of saturation works out to a contribution of $\sim$43 kHz. On a line of width $\Delta\nu \sim 200$ kHz, typical for those operating conditions, that would be about 5% rms sum addition to the linewidth. This example serves to illustrate the trade-off between Q, source power, pressure and hence linewidth, in the choice of operating conditions. In the example given, this contribution to line broadening would hardly be noticeable. For quantitative work where spectral peak *area* is measured, the effect of power saturation will be less apparent than if *height* were the measurand.

With the availability of more powerful sources at lower MMW frequencies however, P_{in} typically $\sim$tens mW, even allowing for lower spectrometer $Q \sim 10^4$, power broadening or saturation would have a more significant effect on the linewidth and consequently on the shape of quantitative response curves. This would become particularly important if signals from species of very different B_{mn} values were to be compared.

There is also a trade-off between maximum allowable power input and sample pressure. In the example given above, a sample at a ten-fold higher pressure 130 Pa would not show power saturation effects until the power input reached 10 mW. For atmospheric pressure samples power saturation ceases to be significant.

3   Line Intensities for Diatomic and Linear Polyatomic Molecules

For diatomic and linear polyatomic molecules where the angular momentum of the molecule arises from end-over-end rotation about its centre of gravity, it can be readily demonstrated that the rotational spectral line frequency neglecting centrifugal stretching, is given approximately by

$$\nu_0 = 2B(J + 1) \tag{1.46}$$

and

$$B = h/8\pi^2 I \qquad (1.47)$$

where J is the rotational quantum number, I the moment of inertia of the molecule/kg m^2 and B is the rotational spectral constant, usually expressed in MHz.*

Equation 1.21 can be rearranged for this special case, by replacing $|m|\mu|n|^2$ with $|J|\mu|J+1|^2$, which is the squared matrix element of that transition, μ being the dipole moment of the molecule. Equation 1.22 can be made more general by adding f_v, the fraction of molecules in that vibrational state, and for completeness taking into account the isotopic fractional abundance i of the molecular species. The resulting formula becomes for all practical purposes (ref. 3, p. 117)

$$\alpha_{max} = \frac{\pi h N f_v f_J i \mu^2 v_0^3}{3\varepsilon_0 c (kT)^2 \Delta v} \qquad (1.48)$$

Kolbe and Leskovar,[7] for example, have calculated the values of α_{max} over a range of J values for a variety of molecules to be found in the atmosphere (*e.g.* Figure 1.1). As another example, at 78 K for CO the maximum theoretical value of α_{max} is 32 m^{-1} and occurs at about 750 GHz, whereas at 300 K the values are for $\alpha_{max} \sim 16$ m^{-1} and $\sim$1500 GHz respectively.

The value of J for the strongest line J_{opt} is given approximately by

$$J_{opt} \approx \left(\frac{3kT}{2B}\right)^{1/2} \qquad (1.49)$$

and

$$v_{strongest\ line} = 2B(J_{opt} + 1) \qquad (1.50)$$

Recalling that $k \approx 20$ GHz K^{-1}, the frequency of the strongest line of a molecule can be readily calculated when B is known.

4 Line Intensities for Symmetric-Top and Asymmetric-Top Molecules

Line Spectra

As the molecules become more complex, so the terms that influence the spectrum and the line intensities increase also although the general form of the relationships does not change radically. For a symmetric-top molecule, one can formulate an approximate expression, adapted from Gordy and Cook (ref. 3, p. 209):

---

*In some texts[3] I is expressed in atomic mass units (amu) and Ångström units (1 Å = 10^{-10} m); whence $BI = h/8\pi^2 = 505\,376$ amu Å^2 MHz.

$$\alpha_{max} \approx \frac{\pi h N f_v f_J i \mu^2 v_0^3}{3\varepsilon_0 c(kT)^2 \Delta v}\left[1 - \frac{K^2}{(J+1)^2}\right] \tag{1.51}$$

where K is the quantum number of the component of the angular momentum along the molecular axis; $|K| \leqslant J$.

As for linear molecules, the frequency of the most intense line can be predicted approximately by Equation 1.50. The intensities for other K states are combined at higher pressures (hundreds Pa) and the frequency of the most intense combination becomes

$$v_{opt} \approx 2B + 12.5(BT)^{1/2}/\text{GHz} \tag{1.52}$$

The values of B for many molecules in the molecular weight range 20–200 Dalton indicate that the optimum spectral region in which to work at ambient temperature is 150–750 GHz.

The situation with asymmetric top rotors is rather more complex because of the wide range of transition selection rules followed, although similar comments apply. The more intense absorption lines will tend to occur at higher frequencies although the relative intensities of lines may vary because of symmetry considerations. Symmetric top spectra occur in *clumps* centred around $2B(J+1)$ and so from the quantitative analysis viewpoint differ hardly at all from those of linear molecules. Asymmetric top spectra are more scattered and it is rather easier to choose an accessible and discrete line from them.

Band Spectra

The majority of polyatomic molecular rotational spectra arise from $J \pm 1 \leftarrow J$, R and P branch transitions and for these, even when they follow asymmetric rotor selection rules, the intensities still increase with increasing frequency: see for example SO_2 (Figure 1.1). A conspicuous exception to this general rule however occurs when Q-branch $J \leftarrow J$ vibration–rotation transitions are present in the spectral range under investigation.

One example of such a band spectrum occurs for the well-documented[2] inversion band in NH_3. This corresponds to a complete inversion of the molecule through its planar configuration, leading to a scatter of rotational lines around the inversion frequency 23.786 GHz. Their frequencies are closely similar, and so their intensities are governed entirely by the J degeneracy, the Boltzmann occupancy factor, the nuclear spin statistical weighting (ref. 2, p. 72) and their transition strengths $K^2/J(J+1)$. These combine to make the 3,3 line at 23.870 GHz the strongest of the band, with lines of other K and J values falling off in intensity at either side.

A second example is that of O_2, in which a magnetic *electron spin-flip* transition occurs at 59.501 GHz. That line is surrounded by a scatter of others due to rotational transitions with their peak intensity occurring at 60.306 GHz. This peak occurs at a higher value of $J = 5$ because O_2 possesses a somewhat lower rotational constant than NH_3.

As a final example, the internal rotor CH_3OH shows not one but a whole series of band spectra arising from molecular tunnelling through the barrier to internal rotation. These are spread through both centimetre and MMW bands, and have been extensively studied, *e.g.* by Xu and Lovas,[8] following on initial work by Lees and Baker.[9]

Many other *floppy* molecules possess band spectra of a similar nature that are well documented in spectral tables. One advantage to the chemical analyst of using these spectra when they exist is that as the pressure rises the individual lines merge into a single broad band with a greater attenuation than that of any one line alone.[10] If interfering species are absent, these provide a sensitive and convenient way of carrying out atmospheric pressure analytical spectrometry (Section 6.9).

5 Significance of the Peak Absorption Coefficient Functions for Quantitative Millimetre Wavelength Spectrometry

For the quantitative analyst, the equations derived above bring with them a reasonably clear guideline. Working at higher frequencies gives access to lines of greater absorption coefficient for a given molecule, no matter what its shape or symmetry.

The optimum frequency at which to work is going to be dictated by other considerations as well, availability of a suitable source and detector, signal to noise considerations, working temperature and pressure, and the other species to be determined. Bringing cost into the equation does focus the mind on the expense of both sources and detectors and unless there were a particular operational need the spectral region 40–200 GHz would be a likely compromise for most applications (see Section 6.2).

Only for particular molecules, *e.g.* ammonia because of its strong lines in the 20–40 GHz region, or water at 22 GHz because there is no other line until 183 GHz, would spectral considerations force the worker to lower frequencies. The 20–40 GHz band is also attractive, however, because of the cheap sources and low-noise semiconductor detectors, manufactured for movement sensors and short-path wireless links. The projected automobile collision-avoidance radar systems will make cheaper sources and detectors available for the 60–70 GHz region within the next few years. The 60 GHz across-office circuits for wireless data links could provide useful narrow-band sources for oxygen determination. The 35 GHz and 94 GHz close-range radar bands provide a useful reservoir of components and sources for the potential manufacturer of MMW spectrometers.

One feature of the inheritance that quantitative MMW spectrometry has from the rotational spectroscopists is the wealth of models, data, spectral constants, tabulated frequencies and line strengths. There are published in the spectroscopy literature extensive tables of spectral data and line strengths for many molecules, which permit in principle the calculation of their α_{max} with sufficient accuracy to predict quantitative analytical performance prior to embarking on measurements.

The single most comprehensive archive is the *Journal of Physical and Chemical Reference Data*. Many of the smaller molecules of pollution concern have also been of interest to astrophysicists,[8] and their journals often hold valuable spectroscopic data. There are, however, myriad other sources of information that can be accessed through electronic retrieval and library systems. Interest in MMW spectroscopy is steady and over the past two decades there have been about four papers a week appearing in the journals, many giving spectroscopic constants for molecules of current interest, with others reporting developments in spectrometer design.

References

1. R. Varma and L.W. Hrubesh, *Chemical Analysis by Microwave Rotational Spectroscopy*, in *Chemical Analysis*, eds. P.J. Elving, J.D. Winefordner and I.M. Kolthoff, Vol. 52, John Wiley and Sons, New York, 1979.
2. C.G. Townes and A.L. Schawlow, *Microwave Spectroscopy*, Dover Publications, New York, 1975.
3. W. Gordy and R.L. Cook, *Microwave Molecular Spectra*, in *Techniques of Chemistry*, ed. A. Weissberger, Vol. 18, John Wiley and Sons, New York, 1984.
4. P.W. Atkins, *Physical Chemistry*, 6th Edn., Oxford University Press, 1998.
5. J.H. Carpenter, A.D. Walters, N.J. Bowring and J.G. Baker, Orientation of Electric Field Tensor in PCl_3 by MMW and Supersonic Jet Spectroscopy, *J. Mol. Spectrosc.*, 1988, **131**, 77–88.
6. J.L. McHale, *Molecular Spectroscopy*, Prentice Hall, New Jersey, 1999.
7. W.F. Kolbe and B. Leskovar, MMW and sub-MMW Absorption by Atmospheric Constituents and Pollutants, *J. Quant. Spectrosc. Radiat. Transfer*, 1983, **30**, 763–478.
8. Li Hong Xu and F.J. Lovas, Microwave Spectra of Molecules of Astrophysical Interest, XXIV Methanol $^{13}CH_3OH$, *J. Phys. Chem. Ref. Data*, 1997, **26**, 17–156.
9. R.M. Lees and J.G. Baker, Torsion-Vibration-Rotation Interactions in Methanol, *J. Chem. Phys.*, 1968, **48**, 5299–5318.
10. H.J. Liebe, Updated Model for MMW Propagation in Moist Air: Data on Oxygen and Water Absorption in the Atmosphere, *Radio Sci.*, 1985, **20**, 1069–1089.

CHAPTER 2

The Components of a MMW Cavity Spectrometer for Quantitative Measurements

The MMW region of the electromagnetic spectrum is characterised by quasi-optical electronic components for generation, transmission and detection. MMW sources and detectors are typically semiconductor devices with transmission lines of open tube rectangular waveguide, or coaxial cable at frequencies below ~30 GHz. The cables are rather *lossy* and waveguide is the favoured transmission line for any other than the shortest links or where dielectric loss is less critical than mechanical flexibility, *e.g.* connecting to a frequency counter.

Commercial MMW components are precisely machined to tight specifications and can be bought on their specification with assured performance. Machining the cavities and interfacing the pumping and sampling systems does not require particularly sophisticated tools, and the mirrors can even be built around surface coated optical reflectors where suitable sizes are available. Experimentation is nonetheless an expensive business for the MMW components alone, particularly at higher frequencies.

This chapter aims to explain the essential features and descriptors of components for a frequency-modulated (FM) source MMW spectrometer so that their specifications can be understood for the spectrometer design. A typical FM spectrometer is shown in Figure 2.1 and the component parts with their interactions are described below.

1 Cavity Absorption Cells

Waveguide cells exhibit a broad bandwidth and consequently find favour in the spectroscopic study of molecules. They are, however, too large for any sort of process analysis or for use on mobile platforms and suffer badly from memory effects due to their high surface area to volume ratio. Whilst not dismissing their importance to the subject in general we have focused on the more compact cavity spectrometers in this work.

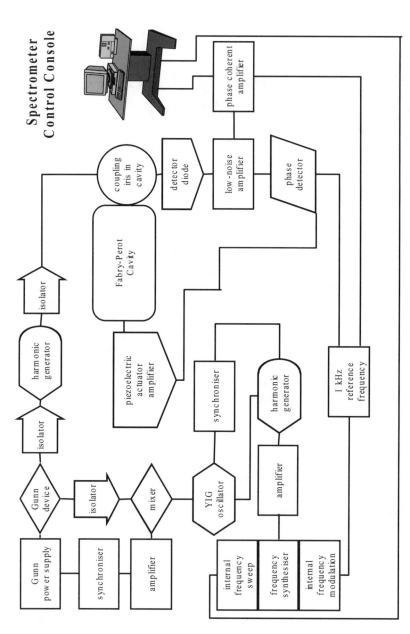

Figure 2.1 *Block diagram of the frequency modulated MMW spectrometer used by the authors. For clarity, not all connections and components are shown. In particular diode and mixer bias power supplies, directional couplers, pumping, pressure measurement and gas blending equipment are absent*
(Redrawn from Rezgui *et al.*[7] with permission from Elsevier Science)

The extinction coefficients of commonly observed rotational spectral lines range from 1 to 10^{-8} m^{-1}. Thus radiation may need to traverse many tens of metres through the sample in order to provide useful absorption signals. In a compact instrument such lengthy traverses are normally achieved by multiple passes of the radiation through an internally reflecting resonant cavity. The preferred form of cavity is an open structure which allows easy access to gaseous samples whilst minimising the absorption of radiation on its reflecting surfaces. Although it is possible in principle to use plane mirrors to realise these multiple passes, the tendency of the collimated beam of radiation to *walk off* their ends after many passes makes alignment critical and sensitivity to external vibration extreme. Instead, a Fabry–Perot structure involving two near-confocal spherical mirrors (Figure 2.2) shows much greater stability and is in widespread use.

A common alternative configuration is the optically equivalent combination of a spherical with a plane mirror, a *semi-confocal* cavity. This is more convenient in that the length of the cavity is reduced by a factor of two, but it does mean doubling the number of reflections and therefore the transmission loss for a given transit path of the radiation through the sample.

The size and shape of the structure are governed by the need to minimise diffraction losses of MMW radiation from the cavity. Stable operation is achievable[1,2] whenever these parameters lie within the unhatched zones displayed in Figure 2.3. The four possible configurations of the mirrors compatible with these conditions are illustrated in Figure 2.4. At the concentric limit characterised by $L = 2R$, the trapped MMW beam is almost conical (Figure 2.4a), and has zero width at the midpoint between the mirrors. As L decreases, the beam becomes paraboloidal in form (Figure 2.4b) and increases in width at the midpoint, the waist. At the confocal point $L = R$ (Figure 2.4c), the radius of the waist at the midpoint reaches a maximum value of $(\lambda L / 2\pi)^{\frac{1}{2}}$. Any further decrease in length

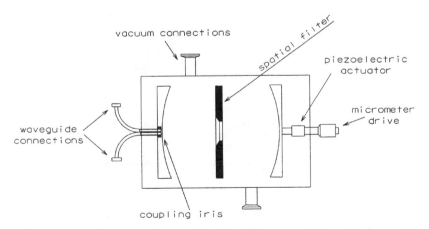

Figure 2.2   *Schematic layout of the Fabry–Perot cavity spectrometer. Vacuum seals and return springs for the micrometer drive and electrical connections for the piezoelectric actuator are not shown*

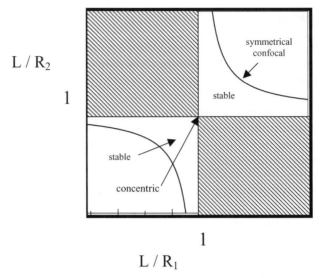

Figure 2.3 *Stability criterion for open optical resonators. The curves plot the ratio of mirror spacing L to radius of curvature R of the two mirrors making up the cavity; in this case $R_1 = R_2$. In the hatched sectors and inside the curve boundaries the stability criterion is not met (see text) resulting in increased loss from the cavityand reduced Q*
(Adapted from Yariv[1] and Pozar[2])

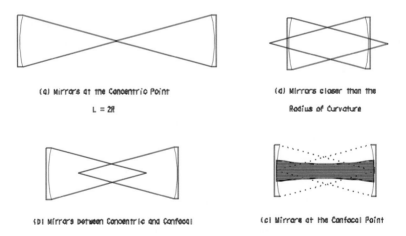

Figure 2.4 *Configurations for a Fabry–Perot optical cavity. The radii indicate the focal point of the mirrors. In (c) the hatched area indicates the beam shape in the confocal configuration (see text)*

leads to a shift of the focal point outside the mirrors (Figure 2.4d) although the beam itself remains confined within them, and to a decrease in the waist radius.

At the same time, the dimensions of the beam also undergo major changes when it strikes the mirrors. At the confocal point (Figure 2.4c) its area is no more than twice the waist area, but as L changes in either direction this area increases rapidly to reach rather large values at the concentric- or the zero-spacing limit. Consequently a confocal configuration is much to be preferred, not only because it permits the use of relatively small mirrors without incurring severe diffraction losses, but because it confines the sampled volume to a cylinder of length L and approximate radius only a little larger than $(\lambda L/2\pi)^{\frac{1}{2}}$. This permits insertion of other components between the mirrors without perturbing the MMW field distribution in the cavity. A cell of length 500 mm and mirror diameter 100 mm, which probably represents the upper limit of size for any transportable cavity design may readily be constructed for use at 100 GHz and above.

The second important design parameter concerns the reflectivity of the mirrors used. Once diffraction losses have been minimised, the beam simply travels back and forth between the mirrors whilst losing some intensity at each reflection. Mirrors can readily be machined and subsequently plated to produce reflection coefficients as high as 99.9% in the MMW region. Silver gives the lowest resistive losses but gold plating tends to be chemically more resistant and would probably give a better lifetime in most applications.

In the example given above such a reflection coefficient would yield a transit distance of 500 m before the beam became attenuated by a factor 1/e. That distance would be traversed in $\sim$1.67 μs, corresponding to a cavity-Q >10^6 at 100 GHz. Such a high-Q would mean that the cavity bandwidth over which it could respond to any input would be in that case $\sim$100 kHz. Some degree of frequency stabilisation of the MMW source would consequently be called for, to ensure it remained within this bandwidth for the duration of any measurement (Section 3.3).

The confocal Fabry–Perot cavity absorption cell displays multiple resonances. The most prominent of these resonant modes are labelled TEM$_{00q}$ and correspond to a smooth Gaussian decay in the lateral radial direction, plus a set of maxima labelled by the subscript q and spaced at half-wavelength $\lambda/2$ intervals in the longitudinal direction. Because the mode number q may take on any integer value consistent with the formation of a standing wave made up of $q\lambda/2$ minima between the mirrors, there will be a large number of possible resonant frequencies. The frequency spacing between these is termed the *free spectral range*, and given by the expression $\Delta v = c/2L$. In the example above it would be 300 MHz, and would result in the observation of resonant peaks with this spacing if a wideband frequency swept MMW source were transmitted through the cavity (Figure 2.5a).

There are in addition, further modes in which the lateral MMW field also varies. These are usually less prominent because they are not so readily excited by the mechanism that couples power into the cavity. They are described by the label TE$_{mnq}$; m and n are further small integers. The general expression describing their resonant frequencies is:

$$v = qc/2L + (m + n + 1)(c/2\pi L)\arccos\{[1 - (L/R)]\}^2 \qquad (2.1)$$

For a confocal cavity the arccos term takes on the value $\pi/2$ and so the lateral

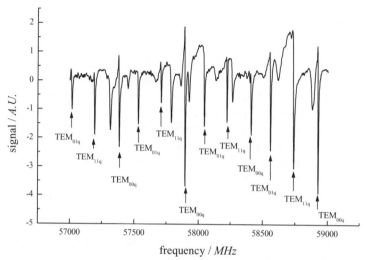

Figure 2.5a   *Wideband frequency sweep of a Fabry–Perot cavity without a spatial filter. The evenly spaced resonances are noted and their identities assigned. In this scan the mode number q ranges from 111 to 114*

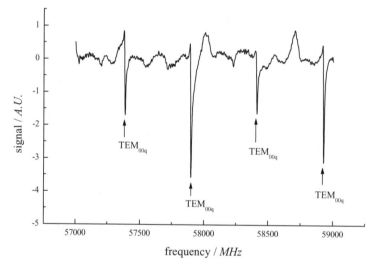

Figure 2.5b   *Wideband frequency sweep of a Fabry–Perot cavity with a spatial filter in place. Note how the non-TEM_{00q} modes are suppressed. In this scan the mode number q ranges from 111 to 114*

modes are spaced by a frequency $c/4L$. In this idealised case they would appear at frequencies both coincident with the longitudinal modes and halfway between them. In practice, however, the arccos argument is only zero at one specific length and the effect of the input coupler is to shift further or suppress some lateral modes rather than others. As a result these extra modes are clearly visible, but appear at less regular positions than predicted by the equation above (Figure 2.5a). They may be eliminated by careful coupler design and by the insertion within the cavity of a lossy cylindrical *spatial filter*[3] with a central hole sufficiently narrow to discriminate against all but TEM_{00q} modes (Figure 2.5b).

2 Analytical Absorption Signals in a Fabry–Perot Cavity

The effect of admitting an absorbing gaseous sample into a Fabry–Perot, or any resonant cavity, is to lower its Q by an amount proportional to the absorption coefficient α defined as the power attenuation per unit length traversed. Thus the incident power P_0 is reduced to P:

$$P = P_0 \exp(-\alpha L) = P_0 \exp(-\alpha c \tau) \tag{2.2}$$

where τ is the signal decay time constant and c is the speed of light in the cavity. The sample causes the power in the cavity to decay with a time constant $\tau_s = 1/\alpha c$. When combined with the natural decay rate $1/\tau_c$ due to the inherent cavity-Q Q_c, the nett decay rate τ of power within the cavity is given by:

$$1/\tau = 1/\tau_s + 1/\tau_c = \alpha c + \omega/Q_c \tag{2.3}$$

So, defining a sample-Q Q_s by:

$$1/Q = 1/Q_s + 1/Q_c \tag{2.4}$$

where the unsubscripted-Q is the combined cavity-Q and sample-Q we obtain:

$$1/Q_s = \alpha c/\omega = \alpha \lambda/2\pi \tag{2.5}$$

The effective path length traversed through the sample before its absorption causes the signal to decay by a factor of $1/e$ is therefore $Q_s \lambda/2\pi$. For a weak line $\alpha \sim 10^{-8}$ m^{-1} observed under the conditions cited above Q_s takes on a value $\sim 2 \times 10^{11}$ and Q_c changes by only ~ 5 ppm. Thus we are seeking to observe rather small changes in the cavity-Q when a sample is inserted.

The effect of the sample is not only to increase the cavity loss function but also to broaden its frequency response. Both these cause a modification of the signal observed by an FM spectrometer and will be considered in more detail in Section 6.8.

3 Coupling Radiation into and out of Absorption Cells

Although this topic is one of considerable importance to the analytical spectroscopist, the optimisation of sample signals generated within a cavity has received little or no attention in most communication systems design textbooks and manuals. These tend to concentrate on optimising power transmission and propagation over narrow frequency bands.[4] In spectroscopic applications, however, we are seeking to optimise the weak signal generated by a sample rather than the transmitted power itself, and in many cases these needs conflict rather than coincide. We feel therefore that a more detailed consideration of optimisation of the cavity plus sample response is justified. It must be admitted too, that there are almost as many alternative coupling configurations used as there are published designs for MMW spectrometers. Some examples of systems described in the literature are shown in Figure 2.6.

It has been shown above (Section 2.1) that the effect of an absorbing sample is to lower the cavity-Q at the sample resonance frequency. This may be observed as a change in the MMW power reflected by the cavity, or by a change in the amplitude of its response to frequency modulation. In small-signal designs such changes are usually monitored by mixing them at a detector with another power source in order to use homodyne or superheterodyne detection;* for a fuller discussion of the principles involved, see Section 3.5. The parameter to be detected is consequently the product of the *source* MMW voltage change produced by the sample, and an externally applied MMW *local oscillator* voltage. Two examples of such detection and its optimisation are discussed further in detail. One example describes a *reflection cavity* in which a signal is reflected by

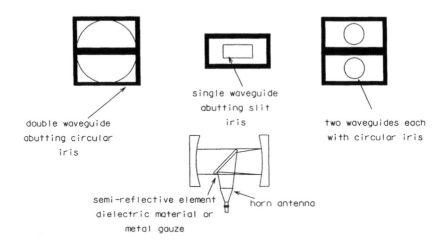

Figure 2.6   *Different coupling schemes that have been used. Each has its particular advantages in complexity, cavity loading and need for other components*

---

*The term *superheterodyne* is an amalgamation of *heterodyne*, referring to the beating of two different-frequency signals, and *supersonic* referring to the frequency of the beat. That beat frequency is usually called an *intermediate frequency* (IF), to which the subsequent filtering and amplification chain is tuned. The IF still carries the audio- or supersonic range *video* modulation envelope. *Homodyne* refers to beating the carrier (or IF) with a locally generated signal of the same frequency, yielding directly the audio or video modulation envelope.

the cavity into the same waveguide as that *via* which the exciting power arrived. The second treats a *transmission cavity* in which the cavity modifies the power passing through it.

It is much simpler to visualise such systems as lumped constant AC circuits rather than by their true description as distributed parameter transmission lines. It must be realised, however, that unless the line or waveguide is terminated in a load that matches its characteristic impedance, power will be reflected from as well as absorbed by the termination. In practical systems, this reflected power may be absorbed by an isolator, mounted in the line in order to avoid a standing wave forming between source and load (Section 3.2). The significant difference between the lumped and the transmission system is that the former operates at fixed internal voltage or current, to deliver whatever power the load can absorb. The latter conversely operates at a fixed power, some of which will be reflected by the load if this is unmatched. When this reflected power is absorbed by an isolator the two become equivalent. A reflection still represents a waste of the input power, however, if it never reaches the sample within the cavity, and most optimised designs seek to minimise such reflections as far as possible. The following treatment describes the criteria for optimisation of the coupling between the MMW source oscillator and the cavity.

The Reflection Cavity

The MMW electrical properties of the cavity and the oscillator at resonance are effectively fixed in their manufacture. This treatment addresses the coupling between those two circuit elements to achieve optimal performance. It considers only the circuits at resonance; that is, under the working condition of the spectrometer. Recall that the capacitive and inductive reactances in a tuned circuit at resonance cancel each other yielding a purely resistive circuit element. Nonetheless the reactances both still remain and are manifest in the property Q. The treatment will yield the reflection coefficient ρ of the coupling interface between the cavity and oscillator in terms of their Qs and the mutual inductance coupling coefficient M of the impedance transformer that the interface represents. Optimum performance will be when the two circuit elements are *critically coupled* to each other and that will be shown to occur when $\rho = 0$.

An appropriate lumped constant[†] circuit diagram to describe this system (Figure 2.7) consists of an AC voltage source V of angular frequency ω and internal resistance Z_0 representing the MMW oscillator circuit, driving a transformer whose secondary is linked to a parallel resonant circuit of resistance R_c symbolising the cavity. The impedance of the secondary circuit as a whole is given the symbol Z_c and the mutual inductance coupling to it by the transformer is given the symbol M. The circuits' respective Qs are:

[†]In these high frequency circuits the inductance, resistance and capacitance are distributed throughout the structure. To facilitate interpretation the effects are *lumped* together and described as discrete components.

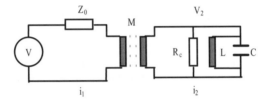

Figure 2.7   *Lumped circuit diagram for a reflection cavity*

$$Q_0 = Z_0/\omega L_0 \text{ and } Q_c = R_c/\omega L_c \qquad (2.6)$$

ωL_c and ωL_0 are the inductive reactance of the cavity and generator circuits respectively; *do not confuse **L** the inductance with L the cavity mirror spacing.*
    We may describe the behaviour of this system by two coupled equations:

$$V = i_1 + j\omega M i_2 \qquad (2.7)$$

$$0 = j\omega M i_1 + i_2 R_c \qquad (2.8)$$

Solving first for the source current i_1:

$$i_1 = V/[Z_0 + (\omega M)^2/R_c] \qquad (2.9)$$

shows that the secondary circuit reflects back a series load equivalent to $(\omega M)^2/R_c$ into the primary. Thus the smaller the cavity resistance and the lower the cavity-Q, the more it loads the primary circuit.
    We can now set down the behaviour of the secondary current:

$$i_2 = -j\omega M i_1/Z_c = -Vj\omega M/[Z_0 R_c + (\omega M)^2] \qquad (2.10)$$

At resonance, the power dissipated in the cavity is given by:

$$P_{cav} = |i_2|^2 R_c = V^2(\omega M)^2 R_c/[Z_0 R_c + (\omega M)^2]^2 \qquad (2.11)$$

Equation 2.11, whether treated as a function of ωM or of R_c, has a maximum when $Z_0 R_c = (\omega M)^2$, and this maximum has the value $V^2/4Z_0$. This is the maximum power available from the source and is delivered to the cavity when the coupling is adjusted to meet this matching criterion.
    Consider now the *transmission line* behaviour of the system; the voltage reflection coefficient ρ of the cavity is given generally by:

$$\rho = (Z_c - Z_0)/(Z_c + Z_0) \qquad (2.12)$$

and in the presently considered condition by:

$$\rho = [(\omega M)^2/R_c - Z_0]/[(\omega M)^2/R_c + Z_0] \qquad (2.13)$$

To simplify the formalism the parameter k can be introduced:

$$k = M/(L_0 L_c)^{\frac{1}{2}} \tag{2.14}$$

It is the coefficient of coupling between the MMW source transmission line and the cavity. This parameter, which includes the only practicable variable M, is the one that needs to be adjusted experimentally for optimum performance. Substitution with k and then Q yields

$$\rho = (k^2 \omega^2 L_0 L_c / Z_0 R_c - 1)/(k^2 \omega^2 L_0 L_c / Z_0 R_c + 1) \tag{2.15}$$

$$\rho = (k^2 / Q_0 Q_c - 1)/(k^2 / Q_0 Q_c + 1) \tag{2.16}$$

The maximum power equation derived above is equivalent to $k^2 / Q_0 Q_c = 1$; therefore the condition for maximum power transfer leads to no reflection from the cavity. The cavity is said then to be *critically coupled* to the transmission line.

If $(\omega M)^2$ is made less than $Z_0 R_c$ the cavity is said to be *undercoupled*; the reflected resistance is less than the optimum value and the cavity will reflect a voltage out of phase with the incident signal. If greater, it is said to be *overcoupled*; the reflected resistance is greater than optimum and the cavity will reflect a voltage in phase with the incident signal. The latter configuration produces a broadening and eventually a splitting of the resonant response, and so meets little favour in practical systems.

Next we may consider the sensitivity of the system to insertion of an analytical sample. This will act to increase the cavity loss, and so lower its Q (Equation 2.4). The optimum sensitivity will therefore occur when the derivative of the appropriate expression above with respect to k and to $1/Q_c$ is a maximum. Choosing this appropriate expression is not, however, quite so straightforward as it may seem. For, with a reflection cavity we are observing not the signal inside the cavity, but that reflected from it. The most appropriate parameter to consider is therefore the derivative of the voltage reflection coefficient ρ with respect to $1/Q_c$:

$$\mathrm{d}\rho/\mathrm{d}(1/Q_0) = 2(k^2/Q_0)/[(k^2/Q_0 Q_c) + 1]^2$$

$$= (Q_c/2)(4k^2/Q_0 Q_c)/[(k^2/Q_0 Q_c) + 1]^2 \tag{2.17}$$

The voltage reflection coefficient and the function in Equation 2.17 multiplying $[Q_c/2]$, the voltage sensitivity, are plotted as a function of $k^2/Q_0 Q_c$ in Figure 2.8a. The voltage sensitivity shows a maximum value of unity at the critical coupling value $k^2/Q_0 Q_c = 1$, but falls off only slowly on each side, reaching its -3 dB value when $k^2/Q_0 Q_c \sim 0.3$.

From Equation 2.1 the maximum voltage signal due to absorption by the sample is $\alpha Q_c/4\pi$, equivalent to a power fluctuation of $\alpha Q_c/2\pi$. It is revealing to compare this signal with that generated by power reflection from the cavity ρ^2 (Figure 2.8b).

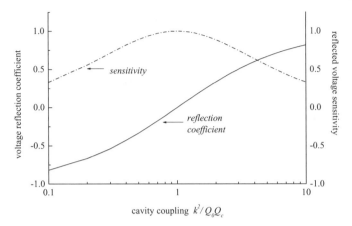

Figure 2.8a *Voltage reflection coefficient,* ——————, *and voltage sensitivity,* – · – · – · –, *to a sample, for a reflection cavity*

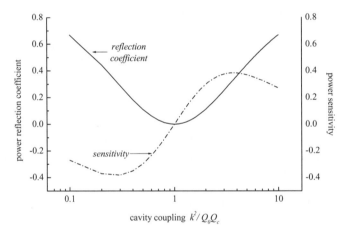

Figure 2.8b *Power reflection coefficient,* ——————, *and power sensitivity,* – · – · – · –, *to a sample, for a reflection cavity*

$$\mathrm{d}\rho^2/\mathrm{d}(1/Q_{\mathrm{c}}) = 2\rho\,\mathrm{d}\rho/\mathrm{d}(1/Q_{\mathrm{c}})$$

$$= 2Q_0(k^2/Q_0Q_{\mathrm{c}})(k^2/Q_0Q_{\mathrm{c}} - 1)/[(k^2/Q_0Q_{\mathrm{c}}) + 1] \qquad (2.18)$$

The power sensitivity exhibits rather different behaviour, showing no signal at all at the critical coupling value and reaching its maximum excursion value of $-0.38Q_0$ when $k^2/Q_0Q_{\mathrm{c}} = 0.27$; that corresponds to $\sim$35% power reflection from the cavity. Such a reflected signal might be useful as an alternative form of homodyne detection. A similar maximum is available at a much larger coupling coefficient, but the overcoupling required makes this a less desirable option.

To summarise the conclusions from this treatment, the optimum signal from a sample in a reflection cavity is obtained when the cavity reflects none of the power incident on it, but absorbs it all. The change in signal due to the presence of a sample then corresponds to an effective traverse of $Q_c\lambda/2\pi$ through it. Ready observation of this change, however, requires some form of homodyne or heterodyne detection and this involves a trade-off. For example, the available power could be split in two by a beam splitter, one half going to the cavity and the other to the detector. The reflected signal must also pass through this beam splitter, with the result that only half of it would arrive at the detector. The nett result would be as if an absorption signal arose from only half of the available incident power.

The Transmission Cavity

In this system the cavity acts as an intermediate coupling device linking two unconnected but identical waveguide transmission lines of impedance Z_0 (Figure 2.9).

Writing down the equations for the currents and voltages at resonance in what are now *three* networks:

$$V = i_1 Z_0 + j\omega M i_2 \tag{2.19}$$

$$0 = j\omega M(i_1 + i_3) + i_2 R_c \tag{2.20}$$

$$0 = j\omega M i_2 + i_3 Z_0 \tag{2.21}$$

leads to:

$$i_1 = V[R_c + (\omega M)^2/Z_0]/[Z_0 R_c + 2(\omega M)^2] \tag{2.22}$$

$$i_2 = -j\omega M V/[Z_0 R_c + 2(\omega M)^2] \tag{2.23}$$

$$i_3 = -V(\omega M)^2/Z_0[Z_0 R_c + 2(\omega M)^2] \tag{2.24}$$

Equation 2.23 shows that two reflected resistances of $(\omega M)^2/Z_0$ now appear in series with R_c in the cavity circuit, whilst an additional load of this value adds to the cavity resistance reflected to the source. The output voltage V_{det} becomes:

$$V_{det} = -V(\omega M)^2/[Z_0 R_c + 2(\omega M)^2] \tag{2.25}$$

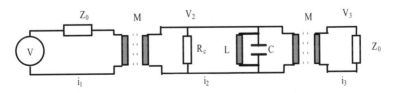

Figure 2.9   *Schematic lumped circuit diagram for a transmission cavity*

whilst the cavity voltage reflection coefficient becomes:

$$\rho = -(\omega M)^2/[Z_0 R_c + 2(\omega M)^2] \tag{2.26}$$

The power dissipated in the cavity and that reaching the detector are respectively:

$$P_{cav} = |i_2|^2 R_c = (V^2/4)[4(\omega M)^2/Z_0 R_c]/[1 + 2(\omega M)^2/Z_0 R_c]^2 \tag{2.27}$$

$$P_{det} = |i_3|^2 Z_0 = (V^2/4)\{4[(\omega M)^2/Z_0 R_c]^2\}/[1 + 2(\omega M)^2 Z_0 R_c]^2 \tag{2.28}$$

Changing these equations to the Q-notation of Section 2.1.3 and calling the power reflection and transmission coefficients respectively ρ^2 and t^2 gives:

$$\rho^2 = 1/(1 + 2k^2/Q_0 Q_c)]^2 \tag{2.29}$$

$$t^2 = 4[(k^2/Q_0 Q_c)/(1 + 2k^2/Q_0 Q_c)]^2 \tag{2.30}$$

The remaining fraction of the incident power $1 - \rho^2 - t^2$ is absorbed by the cavity. Differentiating t^2 with respect to $1/Q_c$ yields the sensitivity to insertion of an analytical sample:

$$dt^2/d(1/Q_c) = 8Q_c[(k^2/Q_0 Q_c)^2/(1 + 2k^2/Q_0 Q_c)]^2 \tag{2.31}$$

The behaviour of the power transmission, reflection and sample sensitivity as a function of the cavity coupling k is plotted in Figure 2.10.

   For this kind of cavity, there is no form of coupling that leads to zero power reflection as characterises the reflection cavity, but the optimum signal sensitivity still occurs at critical coupling. At this point only 11% of the incident power is

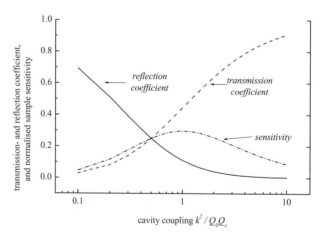

Figure 2.10   *Power transmission, - - - - - - - -, and reflection coefficients, ⎯⎯⎯, and sensitivity, – · – · – · –, to a sample in a transmission cavity*

lost by reflection, the remainder becoming split equally between cavity and detector. Once more, the precise value of the coupling is not so critical, as one can obtain a signal within 3 dB of the optimised value even whilst the coupling coefficient is varied by a factor of 5.

To summarise, an optimised transmission cavity absorbs about half the power incident on it and transmits the rest. Although this may seem at first sight a somewhat poorer performance than that of a reflection cavity, it should be realised that in this configuration no beam splitting components of any kind are required, and that it makes best use of the available power. In the reflecting case further beam splitting components, each with an associated power loss, are needed before the desired signal can be brought to a detector. The relative simplicity of the transmission cavity design has therefore much to commend it.

4   A Practical Coupler

The coupling between the oscillator and cavity may be achieved using a *coupling iris*, which is a small rectangular or circular hole in a septum placed in the waveguide between the two elements. The simplest coupler consists of a circular hole at the centre of a closed feed waveguide. It serves to match the electric field distribution within the waveguide to that within the cavity. We have found that feeding power in and out of a transmission cavity through two parallel open ended waveguides abutting a circular hole of the appropriate dimensions in the cavity wall (Figure 2.6) is often quite adequate to obtain good spectral signals. The coupling coefficient can then be readily modified by the insertion of an aluminium- or brass-foil septum with two suitably smaller diameter holes punched in it.

This approach is not always adequate if wide spectral ranges (several GHz) are required, due to frequency dependent reflections and resonances. The dimensions are so small and the measurements of complex impedance so difficult at MMW frequencies that careful experimentation is usually the only practical way to achieve optimisation. The result has been apertures in the range 2–5 mm wide × 0.2–0.5 mm high, or circular holes of 1–3 mm diameter in brass septa about 100 μm thick for frequencies in the 20–200 GHz region. The criteria for success have been whether the frequency range and sensitivity were adequate and the MMW structure background reasonably flat at the desired operating frequency (Section 6.1).

5   Scanning the Cavity

To measure the spectral line area it is necessary to sweep the source frequency across the absorption line of the gas contained in the cavity. The equivalent path length L_{equiv} in the cavity is given by

$$L_{equiv} = Q\lambda/2\pi \tag{2.32}$$

This would be $\sim$32 m at 150 GHz in a cavity of $Q = 10^5$, that would have a mirror spacing of ~0.5 m as a typical value. At the same time the open structure surrounding the mirrors permits rapid analytical sample insertion and removal, and can be occupied by other vacuum or optical structures without impeding the trajectory of the radiation beam.

One consequence of the high Q attained in these structures is that they become sharply tuned; the system described above would show a FWHM of ~1.5 MHz, comparable with the Doppler width of spectral lines in this region. Thus spectral lines viewed in a cavity may appear as an increased loss that lowers the Q at high pressures[5] whereas at lower pressures their profile becomes distorted because the incident power density varies markedly with offset from the cavity resonant frequency.

A method for overcoming this distortion by mechanically tuning the mirror spacing in synchronism with the tracked MMW source was described in a patent by Leskovar *et al.*[6] and has been developed by us as an analytical technique.[7] The same source FM that is used to detect spectral profiles generates a fundamental frequency component when swept over the cavity profile (Figure 2.11a).

This fundamental component is phase-coherently rectified to give a DC signal which varies from a positive- to a negative-going value as the cavity resonance is traversed (Figure 2.11b). The DC signal is amplified and fed to a piezoelectric actuator that drives a cavity mirror in the appropriate direction to keep the cavity resonant with the source frequency, using a feedback loop to maintain the DC signal close to zero. Thus the peak power transmission of the cavity remains

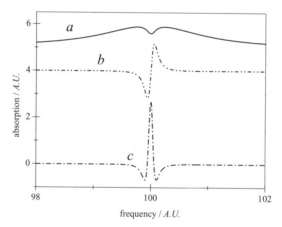

Figure 2.11 *Schematic response to source frequency sweeping of a transmission cavity containing an absorbing gas. **a** represents the cavity with the gas absorption profile superimposed at the cavity resonance. **b** shows the absorption detected at the fundamental frequency and **c** that at the second harmonic of the modulation frequency. The mirror spacing in **b** and **c** would be adjusted to keep the cavity resonance synchronous with the source frequency. The ordinate scale has been offset and exaggerated for clarity, but the abscissa scale is the same for all three traces*
(Redrawn from J.G. Baker, N.D. Rezgui and J.F. Alder, *Anal. Chim. Acta*, 1996, **319**, 277–290, with permission from Elsevier Science)

synchronous with the source frequency, whilst phase coherent detection of the second harmonic of the frequency modulation produces a constant background signal. If, however, the source plus cavity is tracked through a spectral profile the resulting frequency dependent loss appears as a second derivative profile super-posed on this constant background (Figure 2.11c).

A truly constant background would create little difficulty when measuring component concentrations by applying the linewidth or integration algorithms of Section 4.1. The variations generated by the strong coupling and high curvature of the cavity response, however, can become much larger than the signal due to any spectral line. In practice quite small variations in cavity coupling and Q can result in a sloping background as the frequency is scanned. The best way to deal with this in these high-Q cavity spectrometers is by a combination of careful experimental optimisation to minimise the problem, and post-acquisition proces-sing using background correction algorithms (Section 4.3)

References

1. A. Yariv, *Quantum Electronics*, John Wiley and Sons, London, 1975.
2. D.M. Pozar, *Microwave Engineering*, 2nd Edn., John Wiley and Sons, New York, 1998.
3. N.D. Rezgui, J.G. Baker and J.F. Alder, Quantitative MMW Spectrometry (II) Determination of Working Conditions in an Open Fabry–Perot Cavity, *Anal. Chim. Acta*, 1995, **312**, 115–125.
4. M.J. Howes and D.V. Morgan. eds., *Microwave Devices*, John Wiley and Sons, London, 1978.
5. A.F. Krupnov, M.Y. Tretyakov, V.V. Parshin, V.N. Shanin and S.E. Myasnikova, Modern MMW Resonator Spectroscopy of Broad Lines, *J. Mol. Spectrosc.*, 2000, **202**, 107–115.
6. B. Leskovar, H.T. Buscher and W.F. Kolbe, US Patent 4 110 686, August 29, 1978.
7. N.D. Rezgui, J. Allen, J.G. Baker and J.F. Alder, Quantitative MMW Spectrometry (I) Design and Implementation of a Tracked MMW Confocal Fabry–Perot Cavity Spectro-meter for Gas Analysis, *Anal. Chim. Acta*, 1995, **311**, 99–108.

CHAPTER 3

Practical Spectral Sources and Detectors for Analytical Spectrometry

Neither the practising analytical scientist nor indeed many analytical instrument makers would willingly embark upon *ab initio* design and construction of MMW oscillators or detectors. It is much more likely they would purchase the oscillator modules available from the specialist MMW component suppliers. This section aims to help the reader make an informed decision on the suitability of a device for the experiment being planned.

The most commonly encountered MMW sources are now the *Gunn* and *Impatt* devices although the *Backward Wave Oscillator* (BWO) is still used for wideband spectroscopic studies. The Gunn and Impatt devices exhibit the property of *negative resistance* that makes them well suited as MMW oscillators.

1   Negative Resistance Oscillators, Multipliers and BWOs

To understand how a negative resistance oscillator works, it is easiest to consider the electrical characteristics of the waveguide cavity in which the device is housed. A waveguide cavity is in its simplest form a short piece of transmission line terminated in a short-circuit at one end and an impedance matching device coupled to the rest of the circuit at the other end (Section 2.1).

The cavity so formed has both inductive and capacitive properties which, coupled together make the circuit resonant at a particular frequency or frequencies. The cavity also has resistive loss that must be overcome before the circuit can oscillate. In low-frequency circuits this can be achieved with an amplifier whose gain is greater than the resistive attenuation of the circuit. In a negative resistance oscillator the attenuation is overcome by the negative resistance of the diode.

Gunn Devices

Gunn devices belong to a group called *transferred electron oscillators* and are the ones most often encountered in MMW spectrometry, as they offer the lowest noise figure.[1,2] They rely on a bulk property of gallium arsenide and indium phosphide when a DC voltage is applied across the end contacts of the *n*-type material. As the voltage is increased, the current initially increases linearly and then starts to oscillate, with a period closely related to the transit time of the carriers between the contacts across the bulk material. The device is housed in a cavity coupled to a transmission line and is used as a source of MMW radiation, the frequency of which can be tuned mechanically and electronically.

The Gunn device is a homogeneous, thin slice of GaAs or InP with end caps fitted to the terminal leads. At low applied electric field strengths ($<350 \, kV \, m^{-1}$) the charge carriers in GaAs occupy the lowest energy conduction band and the electrons have a relatively high mobility, giving rise to a certain current density in the material that increases with applied voltage. If the electric field is increased above this value, some electrons transfer to an upper conduction band, where they have a lower mobility. As a consequence the current density in the device falls with increasing applied voltage in this region, thus exhibiting a negative resistance. Eventually, population of the upper band dominates the properties and the current density starts to increase again from the lower starting position.

The current *vs.* voltage characteristic of a Gunn device is shown in Figure 3.1. Above a certain bias voltage, the semiconductor material deviates from the usual Ohm's Law characteristic and exhibits the property of negative resistance, before eventually returning to a positive resistance regime. When biased in that region of

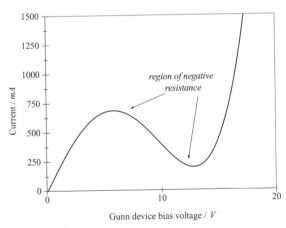

Figure 3.1   *Stylised current vs. voltage curves for a typical Gunn device. The working point would be chosen in the quasi-linear region of negative resistance. The power dissipated is quite large, into a small volume, which is why heat sinking is so important*

negative resistance the device can cancel out the resistive attenuation of the waveguide cavity and the excess negative resistance permits the device and its surrounding structure to oscillate. The voltage amplitude of that oscillation adds to the bias voltage and ultimately its extreme value will push the transient bias into the region where the negative resistance no longer exceeds the resistive loss, and the amplitude of the oscillation becomes maximal. The oscillating cavity is able to transmit its energy into the waveguide transmission line through the coupling iris. This ensures maximum power transfer from the low impedance oscillator, typically a few Ω into the rest of the circuit, that will generally have an impedance of several hundred Ω.

There is no semiconductor junction in a Gunn device. The contacts are, however, separated by only a thin layer of material and heat dissipation is an important design consideration. The thin layer means also that the electric field between the electrodes of a Gunn device is rather high and, whilst otherwise being quite robust, they do suffer from the effects of static electricity when improperly handled. Two typical device packages are shown in Figure 3.2. It is most important to keep the terminals to all MMW semiconductor devices short-circuited during switching operations and when moving the components in and out of circuit. Sources and detectors should preferably be permanently built in to correctly terminated biasing units. Properly looked after, the devices are robust and reliable, with lifetimes measured in thousands of hours. Careful attention to heat-sinking, power supply regulation, switching transient and static discharge suppression and proper training will assure good service from the devices (Section 6.1); one second of inattention and they are dead!

An estimate of the highest frequency of oscillation suggests that GaAs Gunn devices would work up to 70 GHz when in a suitable mount.[2] InP exhibits similar properties to GaAs, and an InP Gunn device should work up to about 160 GHz

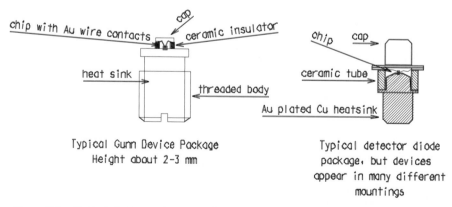

Typical Gunn Device Package
Height about 2–3 mm

Typical detector diode package, but devices appear in many different mountings

Figure 3.2 *Typical device packages. Both are to the same scale. There is no convention as to polarity, so check before you connect!*

but requires an applied field of about 1 MV m^{-1}. This picture is somewhat blurred by the ability to make structures that favour higher harmonics of these oscillation frequencies.

Fundamental mode Gunn sources are available up to 100 GHz with power output >30 mW, giving access with more than enough power to the absorption lines of gases in that region. GaAs devices can produce a few tens of mW at frequencies above 120 GHz. Higher powers would be expected from InP devices, with useful powers available up to 200 GHz. Power output is not generally the critical feature in choosing a practical signal source: a few mW would suffice. The higher power outputs at lower frequencies 50–80 GHz can be used to advantage, however, in driving non-linear devices such as Schottky barrier varactor diodes (described later in this section) to generate higher harmonics, even with only 10–20% efficiency.

Impatt Diodes

Impatt diodes rely on the transit time of electrons in a drift region to create the required negative resistance. These devices comprise a reverse-biased $p - n$ junction followed by a region of semiconductor material, terminated by an n-region. Increasing the applied voltage across the device gives rise to collisional impact ionisation resulting in electrons, which eventually results in an avalanche effect. As the avalanche process takes hold, the electric field starts to collapse and the process slows to virtually nothing. The electron groups are congregated near the $p - n$ junction of the device where the electric field is greatest and after formation the group will start to drift across the region of semiconductor material while the build-up to avalanche repeats itself. The transit of the series of electron groups gives rise to a series of pulses of current in the associated circuitry that forms the Impatt oscillator.

Silicon Impatt diodes are able to yield higher output power than Gunn devices, up to 100 mW at frequencies up to 200 GHz. They suffer from greater sideband noise than do Gunn devices due to the random nature of the avalanche process.[2] The higher working voltage (10–20 V) and a current density of around 100 MA m^{-2} make them rather susceptible also to self-destruction. Careful power regulation and device protection measures when switching are required, as indeed they are for all MMW electronics.

Optimum signal:noise ratio will be obtained at higher frequencies than are available from fundamental mode oscillators, requiring the use of doublers or triplers. Harmonic multipliers are based on semiconductor $p - n$-junction diodes and Schottky barrier devices that are specially designed to show the varactor and step-recovery effects. A semiconductor $p - n$-diode junction behaves like a capacitor when reverse biased due to depletion of carriers near to the junction compared with the concentration in the p and n regions. When a MMW voltage is injected into the junction, the reversing polarity in the second half-cycle gives rise to an asymmetric charging current of the junction capacitance.

This asymmetry in the charging and discharging cycle gives rise to the generation of harmonics of the biasing frequency (Figure 3.3a) and is called

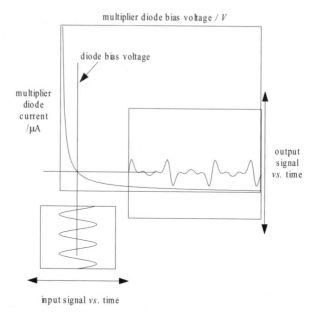

Figure 3.3a   *Stylised representation of harmonic frequency generation in a varactor multiplier diode. The diode is biased in its working region by a small DC current while held in a waveguide or coaxial structure. The impinging radiation is multiplied in frequency by the non-linear characteristics of the current vs. voltage relationship. The harmonic frequencies are then able to propagate in the waveguide*

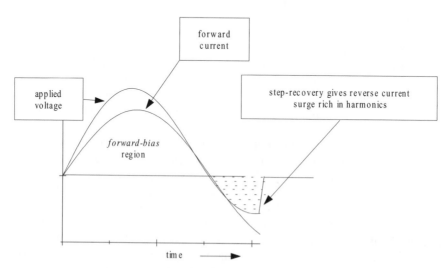

Figure 3.3b   *Schematic representation of the step recovery diode mode of action. During the first part of the forward bias cycle the stored charge is abruptly returned to the junction, producing the step-recovery effect*

the *varactor* effect.[3] Doping levels near the junction can be profiled such that the carriers that have drifted away from the junction during the previous half-cycle, return all-together when the junction becomes forward biased, giving a step-recovery (Figure 3.3b). This abrupt step in the current gives rise to a rich spectrum of harmonics.

Correct construction and tuning of the MMW structure around the step-recovery diode can concentrate the energy in the chosen harmonics, and these devices are used at lower frequencies for both signal and power applications. At the higher MMW frequencies their efficiency falls off, so they are generally used only as harmonic generators with the power spread over all the harmonics, mainly for frequency locking purposes. The Schottky barrier varactor devices are used for multiplication at the higher MMW frequencies where power efficiency is paramount.

Schottky Barrier Devices

Schottky barriers are metal–semiconductor junctions that have the ability to rectify current, because the work function of the metal is greater than that of the semiconductor. The junction thus creates a barrier between the semiconductor and the metal that decreases when the junction is forward biased and *vice versa*. Conduction in Schottky devices is by *majority carriers*, principally electrons. In conventional $p-n$ devices reverse conduction is predominately *via minority carriers*. In $p-n$-junction devices, charge is stored in the junction during forward conduction and has to be removed if the junction is reverse biased before the diode can switch off. The junction capacitance and the capacitive reactance are voltage dependent.

The variable reactance and its related charge storage capacity are used to advantage in some multiplier diodes at lower MMW frequencies, particularly the step-recovery devices described earlier in this section that give frequency multiplication output rich in high harmonics. At higher MMW frequencies however even the small junction capacitance of the step-recovery diode $p-n$ junctions is too great for efficient multiplication. The Schottky barrier varactor devices are therefore employed for higher power efficiency at lower harmonics, typically for doubling or tripling applications.

Schottky barrier varactors do not store charge at all. When driven into forward conduction the Schottky diode absorbs energy and the multiplier efficiency is increased, as long as it is not overdriven to a point where rectification occurs, that is when the MMW power is converted into DC. For a discussion of these devices and their application the text by Maas[4] is recommended. The Schottky barrier varactor multiplier is thus the preferred choice for doublers and triplers at MMW frequencies. Like the other devices they would normally be bought in their waveguide structures with the tuning stubs to match the device impedance to the source and load.

Backward Wave Oscillators, BWO

Although we have focused on solid-state sources for MMW generation, there is still active use of certain vacuum tube electron beam devices, particularly the BWO or *carcinotron* (Chapter 5). The BWO is related to a family of devices that includes also the *Travelling Wave Tube* (TWT) oscillator. An electron beam is caused to travel between an anode and a collector through a periodic metallic structure like a helix or periodically ridged waveguide. The beam induces currents in the metal structure causing a propagating EM wave. The wave can propagate forwards – TWT – or backwards – BWO, depending upon the mechanical structure and electromagnetic environment. The backward propagating EM field in a BWO transfers energy to the electron beam, effectively modulating it, and gives rise to a propagating MMW field that can be coupled into the output waveguide structure.[2,3] The beam needs to be kept in focus using a strong magnetic field, and the beam current is maintained with an anode–collector potential difference ∼ kV.

The BWO is capable of large power outputs and tuning ranges compared with solid state devices. A typical BWO at 100 GHz can produce ∼10 W output. Much more importantly it can be tuned over a wide bandwidth, typically ∼30 GHz and up to 100 GHz.[5,6] It can work up to high MMW frequencies, producing ∼1 mW at ∼1000 GHz.

These characteristics are attractive, particularly their spectroscopic purity, wide tuning range, long lifetime (1–10 kh) and robustness. The drawback is the high beam electrode voltages (4–8 kV) and a magnet for beam focusing, both with their associated power supplies. For analytical equipment this is a significant drawback from the viewpoint of capital and maintenance costs, size, weight and intrinsic safety requirements. This last aspect is not trivial where equipment needs to be installed in chemical plant, confined spaces or on marine and aviation platforms.

Marginal Oscillators

Marginal oscillation is a *state* rather than a device characteristic. It refers to the initial phase in the build-up of oscillation in an electronic circuit. Figure 3.4 shows the typical MMW power output *vs.* bias voltage of a Gunn oscillator exaggerated for clarity. A conventional oscillator would work at V_w; it is possible, however, to work in the marginal oscillation region and this has special character-istics that make it attractive for simple MMW structure designs.

Recalling that a Gunn oscillator comprises the device and an associated resonant structure, the state of oscillation develops when the negative resistance of the Gunn device becomes greater than the positive impedance of the resonant structure to which it is critically coupled (see above). The output power is related to the excess negative resistance of the device over the impedance of the structure. When this excess is low and the negative resistance of the Gunn device is dictated by its bias voltage, the Gunn device can act as a detector, simultaneous with its behaviour as a MMW generator. A change of impedance in the resonant cavity due to the presence of an absorbing gas, for example, will alter the working point of the marginal oscillator, and cause a change in the bias circuit current (Figure 3.4).

This complex relationship means that if the cavity is held at resonance and the spectral line is swept, *e.g.* by Stark or Zeeman modulation, although other modulation schemes are possible, the cavity impedance changes and the reflected power incident on the Gunn device changes in sympathy with the spectral scan. This causes a current to flow in the Gunn oscillator circuit related to the spectral absorption profile, and therefore to its amplitude and area. That current can be readily transformer-coupled out of the Gunn bias circuit and detected synchronously with the modulation frequency.

This simple design has been applied using the equipment described by Thirup *et al.*[7] at the lower MMW frequencies for methanol and ammonia in air. Its full potential has not yet been fully developed for narrow band operation, to which it is best suited. The obvious and critically important feature is the use only of a

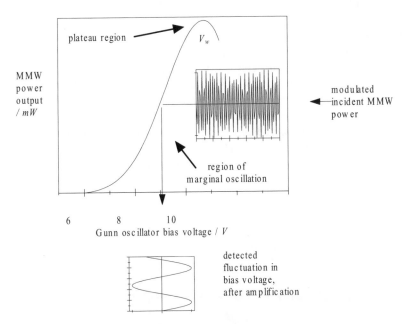

Figure 3.4 *Stylised illustration of the power output vs. bias voltage curve of a Gunn oscillator, demonstrating self-detection of the incident AM MMW signal. The modulation envelope and frequency have been exaggerated for clarity. The region of marginal oscillation is indicated. An oscillator would normally be biased to work in the plateau region where power output was less influenced by bias voltage. The video- or audio-frequency modulation envelope of the detected MMW signal is coupled from the bias circuit and amplified to provide a signal proportional to the amplitude of the incident MMW signal*

single MMW Gunn device as oscillator and detector, and minimal MMW structure.

The work of Dumesh, Surin *et al.*[8,9] describes their *Orotron* devices where power absorbed by the gas is measured through a change in the electron beam current within the structure, when it is in a state analogous to marginal oscillation. The sensitivity and other attractive characteristics of those devices are discussed in Chapter 5.

2   The Use of Isolators in MMW Circuits

It may not always be possible to achieve a stable output at every frequency over the working range of an oscillator, as the complex impedance of the circuit to which it is being coupled will have an influence on its working point. To minimise this problem an isolator is usually placed in between the oscillator and the rest of the circuit to present a constant impedance load to the source.

An isolator allows the transmission of MMW energy through it in the forward direction with little attenuation, typically <1 dB. A signal coming in the opposite direction, arising from a reflection or changing impedance for example, would experience a high attenuation, typically 20–26 dB.

The device is a special case of a *three-port circulator* that relies upon the interaction of a strong magnetic field from a permanent magnet with a ferrite ring, both inside the device.[1] This interaction causes a gyromagnetic motion in the propagating MMW field, causing it to rotate as if going around a traffic roundabout (Figure 3.5a). The forward wave is thus directed to the exit port with little attenuation. A reflection returning to the exit port is also rotated in the same direction and is directed to a third port, terminated in the waveguide characteristic impedance, and is therefore absorbed without reflection. No part of the reflected wave passes to the first, inlet port. The device thus isolates an oscillator from the effect of the external circuit and presents it with nearly constant load impedance.

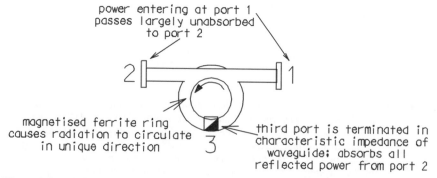

Figure 3.5a   *Schematic representation of an isolator based on a three-port circulator, with one port terminated in the waveguide characteristic impedance*

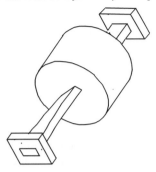

Figure 3.5b   *Typical MMW isolator based on Faraday rotation with twisted waveguide*

At frequencies above 30 GHz particularly, a differently shaped structure is sometimes seen in isolators. These have a characteristic 45° twist in the waveguide termination (Figure 3.5b). They rely on the same phenomenon of Faraday rotation of the polarisation plane that causes this gyromagnetic motion, but it is directed axially rather than horizontally as above. The forward wave is rotated 45°, hence the twist to ensure it passes unimpeded through the waveguide. A reflected wave is twisted first by the waveguide, then further by the magnetised ferrite and so presents itself rotated at 90° to the original input waveguide. There it is below the waveguide cut-off frequency and cannot propagate; instead it is absorbed.

3   Source Frequency Control

Spectral source frequencies need to be controlled to a fraction of a linewidth (Equation 1.27), and this requires some form of frequency locking device, often based on a frequency synthesiser. These are available as units with ranges from low-rf up to >100 GHz, but the cost at the higher frequencies is daunting, and the specification more than is needed for analytical work. Low-frequency synthesisers with ranges up to a few GHz are, however, available at modest cost with resolution of 0.1 Hz and stability better than 1 Hz day^{-1}, well in excess of that required for MMW spectrometry. This is used as a reference oscillator of high stability, to which the spectral source Gunn oscillator is locked. The authors, for example, use a synthesiser with a range up to 1.5 GHz, generally at a frequency around 500 MHz.

To sweep a frequency band the synthesiser generates a series of incremental frequency steps at programmed intervals, say 1000 steps of 50 Hz to the n-th harmonic of which the Gunn source will follow, as explained earlier. As important as frequency stability and resolution is the settling time between steps of the frequency synthesiser and indeed the other parts of the control circuits, which dictates the scan rate and the ability to step between spectral lines quickly. In the example above, a settling time of 1 ms would dictate a sweep time not less than

1 s. Most of all, the synthesiser needs to be brought under external computer control and the programming must therefore be compatible with that used in the rest of the instrument.

Power output of the synthesiser is not critical, but needs to be sufficient to drive a wideband rf power amplifier that will output ~3 W on a 100% duty cycle into a matched load. A peak output of 100 mW from the synthesiser into an rf linear amplifier of about 20 dB gain specified at ~10 W output over the range 100–500 MHz would be adequate.

The amplified rf power is fed to a step recovery diode housed in a microwave structure with output in an intermediate frequency range, typically low GHz. That range is not critical and can be chosen to be compatible with existing or low cost equipment used in other fields, *e.g.* radars and satellite downlinks. The present authors, for example, chose the band 11–15 GHz as it permitted the use of low cost coaxial components, synchroniser and YIG oscillator for the locking steps.

Synchronised YIG Oscillators as Intermediate Frequency Sources

The YIG device is a microwave oscillator that can be tuned in a linear manner over a wide frequency range, and forms the basis of most sweep generators in the MMW region.[3] The principle of operation is similar to a resonant cavity tuned oscillator where an yttrium iron garnet (YIG) is the cavity that can be tuned to resonance by a magnetic field. That field can be modulated, usually with a smaller magnetic field, at frequencies from dc to many MHz (Figure 3.6). The oscillator itself is usually a Gunn device that is freely running in a wideband structure. Its output is coupled to the YIG and the oscillation frequency of the Gunn device is thus dictated by the applied magnetic field. Control of the current through the magnetic coil surrounding the YIG acts to control the oscillator frequency.

In a spectrometer, the spectral source would normally be locked to a frequency

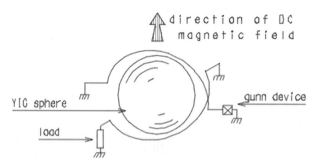

Figure 3.6   *Schematic drawing of a YIG tuned oscillator. When the DC magnetic field is applied, energy is coupled from one loop to the other at the resonant frequency of the YIG sphere, which is itself defined by the magnetic field and can therefore be swept, or held at one frequency*

synthesiser. Signals from the YIG tuned Gunn oscillator and the synthesiser would be compared in a phase sensing circuit, sometimes called a *synchroniser* and used to lock the YIG oscillator to the high-stability synthesised frequency. The synchroniser is an rf version of the more familiar phase-locked loop normally encountered at kHz frequencies. Its heart is a phase-sensitive detector (PSD) that is based on the generation of a difference frequency between the stable output of the frequency synthesiser and the drifting frequency of the YIG oscillator that is to be controlled. This difference frequency or *beat* carries the phase drift characteristics of the YIG oscillator. It is compared in the PSD with a high-stability (hi-stab) local oscillator (LO).[3,4,10] A simplified schematic of a PSD is shown in Figure 3.7; it is similar to the more sophisticated version employed in the Hewlett Packard HP8709A synchroniser used in our work.

The frequency synthesiser is set to give a harmonic frequency that is a predetermined difference from the YIG output frequency. The actual difference is not critical but is a trade-off between a wide pull-in range, loop gain and stability of the synchroniser. The bandwidth of the HP8709A synchroniser was centred on 20 MHz with a pull-in bandwidth of 15–25 MHz. A higher gain, faster synchroniser would permit quicker spectral scanning and settling time as long as it did not introduce instability into the rest of the control circuit. Its mode of action is best understood by considering a typical case, with reference to the block diagram (Figure 3.8).

For example: the desired locked frequency of the YIG oscillator is 12.500 GHz;

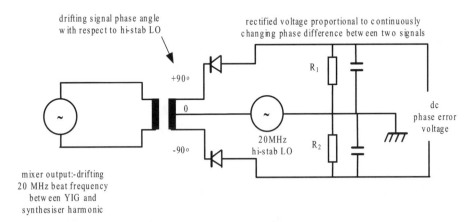

Figure 3.7 *Schematic diagram of a phase-sensitive detector at 20 MHz. The transformer is replaced by an active circuit in the HP8709A synchroniser, and probably most high precision configurations. The phase error voltage output is amplified and used in the spectrometer to control the YIG oscillator magnetic field and hence lock the YIG source frequency to the synthesiser frequency. An identical device locks the Gunn MMW source to the YIG frequency*
(Adapted from Connor[10])

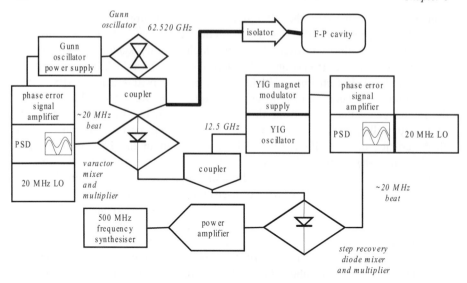

Figure 3.8 *Schematic diagram of the frequency synchroniser and locking arrangement. The first step recovery diode is driven hard and generates harmonics of the 500 MHz synthesised signal. The diode acts also as a mixer for the harmonics and a fraction of the YIG oscillator 12.5 GHz signal, yielding the ~20 MHz beat frequency which is filtered out and passed to the first phase sensitive detector PSD. This is used to lock the YIG oscillator to the synthesiser (see text). The locked YIG output is then mixed in the varactor multiplier diode with a fraction of the 62.520 GHz Gunn oscillator output, yielding the ~20 MHz beat frequency which is passed to the second phase sensitive detector. The Gunn device oscillation frequency is dictated by its bias voltage (see text) and this is used to lock its frequency to the YIG and hence to the synthesiser frequency. The figures in the diagram refer to the example in the text*

the synthesiser is set at the 25th sub-harmonic of 12.520 GHz, 500.8 MHz, and the amplified output used to drive a step-recovery diode, into which is also coupled a fraction of the YIG oscillator output. The step-recovery diode multiplies all the harmonics and the microwave frequency together, but only the frequency near to the 20 MHz beat frequency between the YIG and 25th harmonic signals can pass through the 20 MHz bandpass filter into the synchroniser circuit.

The important point is that the beat frequency carries the phase relationship between the synthesiser and the YIG oscillator outputs. The function of the synchroniser is to reduce that phase difference to zero by sending a control signal to the YIG oscillator to adjust its frequency until that zero phase difference is achieved. At that point the YIG oscillator will be phase-locked to the synthesiser and thus have its characteristic stability and resettability, *viz.* 25×0.1 Hz resolution.

To achieve this there is a hi-stab crystal controlled 20 MHz reference oscillator in the synchroniser circuit that is summed phase-coherently in the mixer with the step recovery diode output containing the $\sim$20 MHz beat to produce the correction signal.

That correction signal has an amplitude proportional to the difference in phase between the reference 20 MHz signal and the $\sim$20 MHz beat frequency. This phase error signal is then amplified and fed back to control the YIG oscillator output frequency. The sign of the phase error signal is such as to force the YIG oscillator frequency towards the synthesiser harmonic frequency. When they become equal the beat frequency from the step recovery diode will be identical to the reference oscillator 20 MHz and the two sources will be locked to zero phase difference.

The next stage is to lock the Gunn oscillator that is the actual spectral source for the measurement at, let us suppose, 62.520 GHz. This is achieved in the same way: the YIG oscillator output is used to drive a varactor multiplier diode held in a MMW structure that couples a fraction of the signal from the Gunn spectral source. The beat frequency at around 20 MHz, between the 5th harmonic of the YIG oscillator at 62.500 GHz and the Gunn oscillator at around 62.520 GHz, is passed to a second synchroniser.

The phase error voltage generated is used to adjust the Gunn oscillator power supply voltage which alters the working point of the Gunn device, particularly its frequency, until the phase error signal becomes zero. In this way the Gunn spectral source takes on the stability characteristics of the synthesiser diluted by the harmonic number, in this case 125. Subsequent multiplication of the Gunn output in a Schottky barrier varactor to give a tripled frequency of 187.560 GHz, would yield an ultimate resolution of 37.5 Hz per synthesiser 0.1 Hz step, which is more than adequate for the measurements required to be undertaken for quantitative analysis.

Frequency-step scanning is thus possible from the synthesiser with the resolution described above. The limit to the scanning rate is usually dictated by the speed and stability of the synchroniser control loops. The scan range is dictated by the locking bandwidth of the synchronisers, and also by the power output of the Gunn spectral source over the bandwidth of the scan. If the scan range is too great, the signal amplitude for the locking circuits may become too small, or the Gunn oscillator may jump to another frequency dictated by the combined properties of the cavity and Gunn source. Over several linewidths this is not usually a problem, but if several lines over a range of hundreds of MHz are to be scanned, retuning of the MMW structures will be required and possibly choice of other sub-harmonic numbers. Human or computer controlled stepper-motor driven tuning elements and frequency measurement of the source output will be required to achieve this.

There are alternative means of source frequency stabilisation;[11] one is to couple an oscillator to a resonant cavity machined from a temperature insensitive alloy as part of a spectrometer, *e.g.* Zhu *et al.*[12] Oscillator modules are available commercially, *e.g.* Farran,[13] Elva-1,[14] that have low phase noise and can be employed as local oscillators in narrow band spectrometer designs. Thirup *et al.*[7]

and Zhu *et al.*,[12] amongst others discussed in Chapter 5, locked the oscillator to the frequency of the rotational spectral line that was being measured. Using rather simple circuitry for long-term stability, at the expense of narrow instrument spectral bandwidth, the spectrometers were aimed at long-term single analyte determination for process monitoring.

Electronic design has overtaken the synchroniser circuits described above and they have been superseded by integrated circuit rf phase-locked loop modules for integration into frequency synthesisers and counters. In an early example of this being applied to spectrometry, Rezgui[15] designed a frequency programmed ~ 22 GHz Stark field modulated cavity spectrometer by comparing its critically coupled Gunn source output frequency to the output of the 20–40 GHz frequency counter used to monitor it. Thus, the desired frequency for the measurement was entered into the controlling computer which adjusted the stepper motor drive controlling the Fabry–Perot cavity until the measured frequency of the Gunn oscillator was near to the programmed frequency. The computer program then fine-tuned the Gunn oscillator until its frequency was the same as programmed. This effectively used the local oscillator and phase locked loops of the frequency counter in a similar way as in the synchroniser-based system described above.

Signal Modulation and Detection Processes

Even after increasing the length of path traversed through the sample by the use of a high-Q cavity, MMW power absorption can still be quite weak, ranging from a few percent to a few ppm of the input signal. Attempting to detect and amplify those absorption signals directly would be difficult because of overloading of the amplifier by unabsorbed power that has passed through the sample. It is possible in principle to cancel out this power using a bridge spectrometer configuration, as is the practice in ESR spectrometry and in dielectric loss measurements[16] discussed in Chapter 5. The balance of such configurations is, however, sensitive to variation in the frequency even over a single linewidth, and they have not found favour in practical MMW spectrometers. Instead, alternative techniques for modulating the spectral signal alone, and then amplifying it synchronously with this modulation, are in universal use. Not only does such a procedure overcome the problem of overload by unabsorbed power, but the modulation frequency can be selected to minimise the non-thermal noise characteristic of most MMW detectors (Section 3.5).

Although the only direct modulation process to be discussed in detail in this monograph is frequency modulation it is by no means the only detection method possible. The most commonly employed method in analytical microwave spectroscopy has been Stark modulation, but as has been discussed above, it is not well suited to cavity MMW spectrometry.

Alternatively one may modulate the sample concentration by periodically removing it from and replacing it within the absorption cell as done, albeit rather slowly, in pulsed supersonic jet absorption spectroscopy.[17] The sample absorption coefficient can be modulated by double resonance using a second microwave source to pump a transition that has one energy level in common with the monitored transition[18] (Figure 1.3). By switching this pump on and off, the population of the common energy level is made to vary, and with it the strength of the monitored transition.

There is also a technique for converting absorption to emission signals and thereby making them readily observable by superheterodyne detection. This is the principle of pulsed FT spectroscopy, in which a short intense polarising pulse of radiation is applied to the sample and the response after it is switched off is measured. The resulting time-varying emission is made up of all the resonant frequencies and absorption strengths displayed by the sample over a finite frequency band corresponding to the bandwidth of the polarising pulse. Allied with the use of a pulsed supersonic jet and computerised summation of the signals generated by repeated pulses it has proved a powerful, if technologically demanding way of observing signals from low concentration samples.[19]

The two constraints that make direct observation of weak absorption signals impracticable are the presence of *pink* noise, which contains a preponderance of low-frequency power compared with white noise, and the enhancement of this and other sources of noise by the rectification process through which MMW signals are detected. Both predicate the use of some kind of modulation at a frequency greater than the noise corner frequency and the use of a phase-coherent detector to convert the resulting modulated signal to a DC level suitable for display or for post-detection computer processing.

Whichever method of modulation is used, the spectral signal from the sample appears superposed on the modulation frequency rather than as a simple plot of absorption *vs.* frequency. To avoid reintroducing noise by rectifying this signal in a final down-conversion to DC, it is customary to extract it by phase-coherent detection, or lock-in amplification. By passing the signal through a gate driven at the modulation frequency, it is possible to generate a rectified signal whilst restricting noise to a narrow frequency band, set by a built-in integrator.

Many of the bandwidth and noise reduction features that were formerly incorporated in electronic hardware can now be replaced by high-speed software; this development has greatly enhanced system performance (Section 4.3). It not only increases the options available to the user, in trading off sensitivity for rapid response or simultaneous observation of several spectra. That capability offers also the possibility of intelligent tracking and search for built-in fingerprint spectral features, opening the possibility of qualitative identification of gaseous mixtures.

A widely used practice is to apply a sinusoidal frequency modulation (FM) to the MMW source. This too produces an amplitude modulation of the transmitted signal. Any frequency dependent component in the MMW transmission line, whether it be a sample, a cavity or a reflecting component, will show similar behaviour. It is important to minimise those spurii by careful construction and attention to mechanical matching of components such as coaxial connectors and waveguide flanges.

A simplistic description of this process for sinusoidal FM of a source output HWHM $\Delta\nu$ at angular frequency ω, is obtained by expanding the frequency varying spectral signal seen at the detector $F(\nu + \Delta\nu \sin \omega t)$ in a Taylor series:

$F(v + \Delta v \sin \omega t)$

$$= F(v) + \Delta v \sin \omega t \, dF(v)/dv + 0.5(\Delta v \sin \omega t)^2 \, d^2 F(v)/dv^2 + \ldots$$

$$= F(v) + \Delta v \sin \omega t \, dF(v)/dv + 0.25(\Delta v)^2 (1 - \cos 2\omega t) d^2 F(v)/dv^2 + \ldots \quad (3.1)$$

Coherent detection of this signal at angular frequency ω will extract the first derivative of the spectral profile, at frequency 2ω the second derivative, and so on (Figure 3.9).

First derivative presentation has the disadvantage that a complete line profile scan is required for sample analysis, because the signal vanishes at the line centre. In contrast the use of the second or other even harmonics for signal detection results in a maximum signal at the line centre, a known frequency that can be programmed into a self-optimising automated system. Furthermore, it is now possible by digital techniques to synthesise modulating waveforms that optimise the signal at particular detection harmonics, and to use digital filtering to select those frequencies at the expense of others (Section 4.2).

The simplest modulation procedure of all involves sweeping the MMW source repeatedly through a spectral line at the chosen frequency, and amplifying the resulting signal with a low-noise high-pass amplifier. Because the noise corner (Section 3.5) of earlier detectors fell in the 100 kHz region, it was not previously practicable to do that whilst simultaneously stabilising the MMW source centre frequency to that of the spectral line during a rapid sweep. In addition the wide bandwidth necessary to reproduce the line profile correctly would make for a poor signal to noise ratio. Recent improvements in control loop technology together with the much improved noise performance of modern Schottky barrier mixers (Section 3.4) have, however, made such sweeps possible, whilst on-line computer signal averaging now deals with the signal to noise problem. Krupnov *et al.*[6]

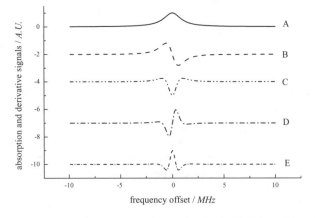

Figure 3.9 *Fundamental* (A), *first-* (B), *second-* (C), *third-* (D) *and fourth-derivative* (E) *of the spectral profile signal*

describe a computer-interfaced instrument in which a complete sweep is taken every 60 μs, enabling collection, averaging and fitting of a spectral profile to be carried out in <1 s (Chapter 5).

4 Detectors for Quantitative MMW Spectrometry

The detector is the most important part of a MMW spectrometer. The photon energy at MMW frequencies is similar to kT (Equation 1.15) and so the noise naturally occurring in the semiconductor diode detectors is significant at these frequencies. The sources are spectrally bright and their close-in sideband noise

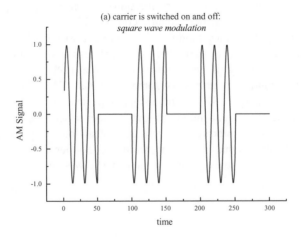

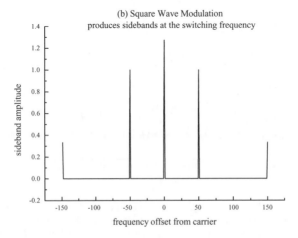

Figure 3.10 *Square wave modulation yields sidebands at intervals of the modulation frequency*

low. The measurement of signals is consequently detector-noise rather than source-noise limited in this spectral region.

This limitation gives rise to the technique of modulating the MMW carrier frequency with a low-frequency waveform and then measuring the analytical absorption signal at that low frequency. When the absorption causes an amplitude modulation of the detected signal, sidebands are produced offset from the MMW carrier at multiples of the modulation frequency (Figure 3.10). The modulation broadens the MMW signal by an amount less than the spectral absorption linewidth, so the magnitude of the modulation sidebands is a true reflection of the absorption of the MMW radiation by the sample. They are extracted from the MMW carrier by heterodyne detection, which is the mixing of signals in a non-linear device to yield the intermodulation products. The signal derived from a Schottky barrier diode detector, a bolometer or any non-linear detector when a modulated MMW signal falls on it is composed of the intermodulation products of the various frequency components of the power spectrum, giving rise to the term *mixer* for these heterodyne detectors (Section 2.1.2).

When the signal is frequency modulated (Figure 3.11a) modulation sidebands still appear, but their phase is such that no signal results on demodulation by heterodyne detection. Only if the cavity or sample itself shows absorption dependent on the source frequency is some of the applied FM converted to amplitude modulation (Figure 3.11b) and a signal detected.

Because the modulation frequency is so low, it is never necessary to carry out a full sideband analysis to interpret the results. Instead one may treat the low-frequency demodulated signal as arising from the Fourier components of the slowly varying response of the system to a tracked MMW source.[20]

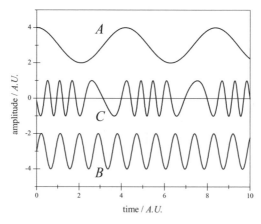

Figure 3.11a   *Stylised illustration of frequency modulation. The low frequency waveform A causes modulation of the carrier frequency B resulting in the frequency modulated waveform C whose frequency varies with time. The magnitude of the carrier frequency change is the deviation and the rate is the frequency of the modulating waveform A*

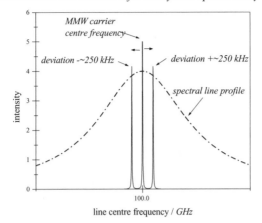

line centre frequency / GHz

Figure 3.11b *Comparison of the spectral linewidth and maximum deviation of the FM source frequency. The frequency deviation at the MMW frequency would be typically ~250 kHz and linewidth HWHM ~1–2 MHz. The carrier sweeps between the deviation limits at the source modulation rate, 1 kHz typically*

Bolometer Mixers

There are two approaches to detection, the best being with a bolometer and the alternative being a semiconductor device. That there should be any contest at all between the two approaches for quantitative analytical spectrometry hinges only on practicality and cost. Lesurf[2] gives an excellent appraisal of these methods and the text is recommended for supplementary reading.

Bolometric or thermal detection methods rely on the heating effect of the MMW radiation. The only bolometer to be discussed here is the liquid helium cooled InSb resistive element device. It alone has the speed of response required for the frequency modulation and phase-sensitive signal recovery techniques employed in quantitative work (Figure 3.12). A crystal of InSb cooled to liquid helium temperatures contains electrons only loosely coupled to the crystal lattice. This so-called *electron gas* is able to move around the lattice quite freely. Incident MMW radiation on the crystal is absorbed by the electron gas, causing the electrons to move around faster. The increased momentum of the electrons gives rise to increased conductance. The electrons are able transiently to maintain a higher temperature than the rest of the crystal because the thermal coupling between the gas and the lattice is so low. The combination of low heat capacity of the electron gas and the low thermal conductivity between it and the lattice results in a bolometer that has a fast response time and high sensitivity.

An InSb bolometer working at the temperature of boiling helium (~4.2 K) will have an intrinsic response time of about 1 μs, a sensitivity of the order 10 V mW^{-1} and a noise equivalent power below 10 pW Hz$^{-\frac{1}{2}}$. The working bandwidth of the detector spans from DC to the IR region, and response is effectively flat over the MMW spectrum. The QMC Instruments Ltd. InSb

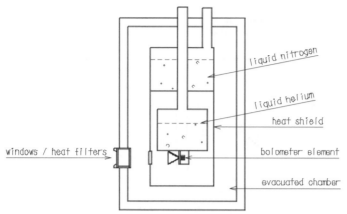

Figure 3.12 *Schematic drawing of the helium-cooled bolometer. Heat filters cut out unwanted radiation and present a cold image to the bolometer*

bolometer used by the present authors had optimised true electrical noise equivalent power of $0.17 \, \mathrm{pW \, Hz^{-\frac{1}{2}}}$ at 270 Hz and an optical signal response of $5.6 \, \mathrm{V \, mW^{-1}}$.

These rather impressive figures are the reasons for working with a helium-cooled bolometer. The response time permits the use of phase-sensitive detection techniques at kHz modulation frequencies. The penalty to be paid is in capital and running costs because of the cryogenic cooling. An InSb bolometer sensor must be operated below 20 K to be useful, and sensitivity continues to increase down to 1 K, with liquid helium temperatures as a practical compromise. As the temperature of the device rises the noise and response sensitivity characteristics degrade, although intrinsic response time does improve to $0.1 \, \mu s$ at 20 K, permitting higher modulation rates and therefore reduced $1/f$ noise. The optimum conditions, however, favour lower temperature operation.

Liquid helium and nitrogen are expensive because supply infrastructure and maintenance costs are considerable, especially if those cryogenic facilities are required only for the bolometer. For many applications and in some parts of the world, their use would not be practical. One is forced therefore to consider other types of detectors as alternatives.

Schottky Diode Mixers

The Schottky diode is the most practically useful and affordable solid state detector at present and can be obtained for frequencies up to 325 GHz. In terms of frequency range and noise performance the nearest contender, the Si point-contact diode mixer, has a noise figure of 11–14 dB over the range 60–80 GHz, its maximum operating frequency. Compared with 5–6 dB for a Schottky device over the same range[1] there is no real contest, even at those relatively low frequencies.

Cryogenically cooled detectors employ the low-noise GaAs Schottky barrier *Mott* diodes.[4] Between 140 and 220 GHz they exhibit 400 K noise equivalent temperature at a lower limit junction temperature of 20 K, below which the performance degrades. The noise temperature is around 1000 K at 300 K junction temperature.[13] Sensitivity of a Schottky barrier mixer diode ranges from about 2.75 V mW^{-1} to 1 V mW^{-1} over the range 90–325 GHz.[13] In comparison the helium-cooled InSb bolometer used by the present authors (Section 3.4.1) can provide double sideband noise temperatures of 200–300 K in the region 100–300 GHz and sensitivity of 5–6 V mW^{-1}.

In an FM MMW spectrometer the spectral source frequency is modulated at a certain rate f, typically $\sim$1 kHz. This gives rise to sidebands of the spectral source frequency above and below the carrier frequency.[18] The frequency modulated MMW carrier has in its modulation envelope phase and amplitude relationships to the carrier. Mixing in the non-linear junction of the detector yields the modulation signals altered by their interaction with the cavity and gas inside it, with their preserved amplitude and phase relationship to the original modulation signals. Those properties are measured by passing the heterodyne mixer output and the thermal noise contribution from the mixer, to a filtered phase-sensitive detection system, with the original modulation as reference.

The phase-sensitive amplifier has a certain response bandwidth and will therefore measure a signal due to the thermal noise from the mixer over that bandwidth, limiting the ultimate signal to noise ratio of the system. The nature of noise in these mixers and detectors is discussed in Section 3.5.

In reading the noise specifications for mixers or detector diodes it is important to determine exactly what is being quoted. The device performance will be influenced by the circuit into which it is incorporated, and its noise figure will depend in part on the way it was measured.

The ratio of the noise power output of a device at any temperature to that of a black body in a chosen standard state is called the *noise figure*. This last term is applied to a variety of devices and may have several different meanings. In the present context it is the ratio of the noise power from a device terminated in its characteristic impedance and biased in its working condition, to the noise power of a pure resistance under the same conditions at the same temperature. It can be written in dB notation or expressed as a *noise temperature* through the relationship *noise figure*/dB $= 10 \log(T_{\mathrm{m}}/290)$. The temperature 290 K is meant to represent a black body source at room temperature and T_{m} is the noise temperature of the device. This form is useful for comparing devices operating near room temperature, although it can lead to confusion if the exact comparison being made is not specified.

N.B. because the device is non-linear its noise figure will change depending upon the current it is passing: hence the need to specify its being biased in its working condition.

When sources are progressively cooled the noise temperature approaches the actual temperature, but cannot equal it. Lesurf[2] demonstrates that for a single sideband receiver the minimum possible noise temperature is given by

$$T_{\min}/K = 0.047\nu/\text{GHz} \tag{3.2}$$

corresponding to $1-10$ K over the range $20-200$ GHz.

The significance of detector or mixer noise figure is its critical influence on the overall noise output of the spectrometer. As the first element in the amplification chain the mixer can be readily demonstrated to have the most important noise contribution of any of the stages. The noise figure and the noise temperature are equally satisfactory as figures of merit for comparing different devices.

5 Noise and Signal Processing in MMW Spectrometers

Because of the relative weakness of MMW spectral signals and the noise characteristics of MMW detectors, considerably more attention needs to be paid to the optimisation of detection systems than is customary in the optical or infrared regions. In particular, spectral lines invariably occur as absorption superposed on a more intense background of unabsorbed power, rather than as fluorescence or induced emissions with little or no background. The techniques for detecting weak signals, such as the homodyne and the superheterodyne commonplace in radio and MMW receivers, are consequently of only limited value. They amplify not only the desired signal but also its background, and at some point the latter must be removed if the receiver is not to be overloaded. Furthermore, the carrier frequency down-conversion process needed to detect these signals itself generates additional noise. The favoured method for detecting these signals is consequently to use some form of low-frequency signal modulation, to rectify the MMW power directly and then to extract the desired signal from the modulation envelope by phase-coherent detection. Once the signal is obtained, it may then be further improved by computerised enhancement and smoothing.

The primary sources of noise in this region are classified as thermal or *Johnson* noise together with *shot* noise due to the particulate nature of photons and electrons. Additionally there is *non-equilibrium* or *inverse frequency* noise, often termed *pink* noise to distinguish it from the other two that have a uniform white noise frequency distribution.

Johnson Noise

Thermal fluctuations are unavoidably present in all circuits and give rise to a

$$\text{mean noise power} = kTB \tag{3.3}$$

where B is the circuit bandwidth/Hz.

This corresponds to a power $\sim 4 \times 10^{-21}$ W Hz^{-1} at 290 K which produces an rms noise voltage $(4kTBR)^{\frac{1}{2}}$ across any resistor combination of value R. The $p-p$ fluctuation is $\sim 4-5$ times greater than this value. Practical circuit devices such as amplifiers increase this apparent noise power by their power gain plus their own noise figure, and the overall effect gives a noise temperature T_n which replaces the temperature T in the mean noise power kTB.

Shot Noise

Shot noise fluctuations are caused by the non-uniform arrival of photons or electrons at a detector. If the average rate of arrival is N the rms fluctuation in this rate is $N^{\frac{1}{2}}$. As the sampling time is reduced by increasing the circuit bandwidth, so the relative fluctuation increases proportionately to this bandwidth. At MMW frequencies photons have such low energies that the photon flux is high at the powers typical of the spectral sources employed ($\sim 10^{20}$ s^{-1}) at 150 GHz, and the fluctuations are small compared with the thermal noise averaged over the same counting period. When the same signal is converted into a current, however, the discrete electronic charge produces a rms fluctuation of $(2eIB)^{\frac{1}{2}}$ in the current I. This variation is comparable with the thermal noise in a circuit as soon as the steady voltage present reaches 50 mV.

Inverse Frequency or $1/f$ Noise

Inverse frequency noise has a power spectrum corresponding to the formula:

$$P_n = CI^2 B/f \tag{3.4}$$

where C is a multiplying constant and f is the detection frequency. It is associated with the random trapping and release of current carriers at the surface of a solid, and is particularly noticeable for the large surface area to volume ratios characteristic of MMW point-contact detectors.[3] It is for this reason that direct signal rectification by such devices cannot be used for effective small-signal detection, and the experimenter must resort to the more complex schemes to be described below. Indeed it is customary to define a *noise corner frequency* for such devices at which white and pink noise contributions are equal, and to make sure that all detection is carried out at a higher frequency than this. In early spectrometers this noise corner frequency was as high as 100 kHz[11] and it was often necessary to resort to some form of rf detection. Modern Schottky diode detectors show noise corners of <1 kHz and as a result the design of low noise signal amplifiers has become much more straightforward.

Signal Rectification Noise

In order to account for excess noise inherent in the rectification process, we may consider a simplified model valid at very low power levels. This implies that the detector is a linear power detector exposed to MMW power P_m or oscillating voltage V_m:

$$V_{out} = aP_m = bV_m{}^2 \tag{3.5}$$

If noise P_n is present on the microwave signal, this too will be rectified:

$$V_{out} + \delta V = b(V_m + V_n)^2 = bV_m^2 + 2bV_mV_n + bV_n^2 \qquad (3.6)$$

When we average the signal over a long time on the both sides of the equation, the terms linear in V_n disappear, leaving:

$$V_{out} = b(V_m^2 + V_n^2) = a(P_m + P_n) \qquad (3.7)$$

so that the signal caused by microwave noise appears simply to add to that of the coherent microwave power. However, we must also consider the fluctuations in V_{out} caused by this additional noise signal, and for these:

$$\delta V^2 = (2bV_mV_n)^2 = 4a^2 P_m P_n \qquad (3.8)$$

Thus the rms noise voltage generated by the fluctuations in power is given by:

$$\delta V_{rms} = 2a(P_m P_n)^{\frac{1}{2}} \qquad (3.9)$$

This signal is no longer linearly dependent on the noise power as was the rectified noise signal above; rather it is *pumped up* by the coherent power to give a signal larger than the former by a factor $2(P_m P_n)^{\frac{1}{2}}$. Simple rectification of a microwave signal leads therefore to an output with an extremely poor signal to noise ratio, because noise signals from the entire microwave band sampled are picked up and enhanced by the rectification process.

In both radio and microwave receivers this noise amplification problem is overcome by the process of superheterodyne detection. The incoming signal is mixed with a more powerful local oscillator at a similar but not identical microwave frequency, generating a beat at the difference frequency between the two. The effect of this mixing in the above equations can be seen, by substituting the expressions P_{LO} and P_m for P_m and P_n respectively.

The output voltage contains a term $2a\sqrt{(P_m P_{LO})}$ at the beat frequency that was amplified in precisely the same manner as was the noise in those equations. Both signal and noise are amplified by the same amount and so the signal to noise ratio at this beat frequency is unaffected by the non-linear mixing process. To be more precise, the signal to noise ratio is degraded by 3 dB because system noise is also detected at the sum of signal and local oscillator frequencies, and by a further 1–2 dB because of non-ideality of the mixing process. This is a small price to pay compared with the orders of magnitude higher degradation that is produced by direct rectification.

Weak spectral absorption is, however, characterised by small changes in a large background power and is therefore ill suited to superheterodyne detection. That background is also detected and overloads the sensitive intermediate frequency amplifier tuned to the beat frequency that is derived from the mixing process. Instead it is customary to modulate the absorption in some way and to observe the modulation on the detected signal rather than the absorption itself.

In Section 3.4 we have shown how modulation processes can in the presence of a sample create amplitude modulated MMW signals that contain sidebands offset by multiples of the modulation frequency (Figures 3.10 and 3.11). On detection these sidebands now act as the signal seen by the carrier acting as a local oscillator and so contribute to a beat frequency signal that is itself at the modulation frequency. So long as this frequency lies above the noise corner of the detector, the only noise contribution to the beat will be white thermal noise.

References

1. T.S. Laverghetta, *Microwaves and Wireless Simplified*, Artech House, London, 1997.
2. J. Lesurf, *Millimetre-Wave Optics, Devices and Systems*, Adam Hilger, IOP Publishing, Bristol, 1990.
3. M.W. Dixon, ed., *Microwave Handbook*, Vols. 1 and 2, Radio Society of Great Britain, 1989.
4. S.A. Maas, *The RF and Microwave Circuit Design Cookbook*, Artech House, London, 1998.
5. A.F. Krupnov, Present State of sub-MMW Spectroscopy at the Nizhnii Novgorod Laboratory, *Spectrochim. Acta, Part A*, 1996, **52**, 967–993.
6. A.F. Krupnov, M.Y. Tretyakov, V.V. Parshin, V.N. Shanin and S.E. Myasnikova, Modern MMW Resonator Spectroscopy of Broad Lines, *J. Mol. Spectrosc.*, 2000, **202**, 107–115.
7. G. Thirup, F. Benmakroha, A. Leontakianakos and J.F. Alder, Analytical Microwave Spectrometer Employing a Gunn Diode Locked to the Rotational Absorption Line, *J. Phys. E: Sci. Instrum.*, 1986, **19**, 823–830.
8. B.S. Dumesh, V.P. Kostromin F.S. Rusin and L.A. Surin, Highly Sensitive MMW Spectrometry based on an Orotron, *Meas. Sci. Technol.*, 1992, **3**, 873–878.
9. B.S. Dumesh, V.D. Gorbatenkov, V.G. Koloshnikov, V.A. Panfilov and L.A. Surin, Application of Highly Sensitive MMW Cavity Spectrometer Based on Orotron for Gas Analysis, *Spectrochim. Acta Part A*, 1997, **55**, 835–844.
10. F.R. Connor, *Introductory Topics in Electronics and Telecommunication: Modulation*, 3rd Edn., Edward Arnold, London, 1985.
11. M.J. Howes and D.V. Morgan, eds., *Microwave Devices*, John Wiley and Sons, London, 1978.
12. Z. Zhu, I.P. Matthews and A.H. Samuel, Design and Implementation of a Cavity Microwave Spectrometer for Ethylene Oxide Monitoring, *Rev. Sci. Instrum.*, 1995, **66**, 4817; *ibid.*, 1996, **67**, 2496; Z. Zhu, I.P. Matthews and G.W. Dickinson, *ibid.*, 1997, **68**, 2883.
13. Farran Technology Ltd., MMW Component Product Catalogue, 2001, http://www.farran.com
14. ELVA-1 MMW Divn., Components and Systems Product Catalogue, 2001 http://www.elva-1.spb.ru
15. N.D. Rezgui, Design of an Automatic Microwave Spectrometer, PhD Thesis, University of Manchester, 1991.
16. J.F. Rouleau, J. Goyette, T.K. Bose and M.F. Frechette, Investigation of a Microwave Differential Cavity Resonator for the Measurement of Humidity in Gases, *Rev. Sci. Instrum.*, 1999, **70**, 3590–3594.
17. J.H. Carpenter, A.D. Walters, N.J. Bowring and J.G. Baker, Orientation of Electric

Field Tensor in PCl₃ by MMW and Supersonic Jet Spectroscopy, *J. Mol. Spectrosc.*, 1988, **131**, 77–88.

18. D.A. Andrews, N.J. Bowring and J.G. Baker, Novel Coherence Effects in Microwave-Microwave Double Resonance Pulse FT Spectroscopy, *J. Phys. B*, 1992, **25**, 667–678.

19. J.U. Grabow, W. Stahl and H. Dreizler, Multi-Octave Oriented Beam Resonator FT Microwave Spectrometer, *Rev. Sci. Instrum.*, 1996, **57**, 4072–4084.

20. J.G. Baker, N.D. Rezgui and J.F. Alder, Quantitative MMW Spectrometry (III) Theory of Spectral Detection and Quantitative Analysis in a MMW Confocal Fabry–Perot Cavity Spectrometer, *Anal. Chim. Acta*, 1996, **319**, 277–290.

CHAPTER 4

The Quantitative Analysis of Gas Mixtures

This chapter brings together the key components described in the previous sections to produce a prescription for a practical spectrometric method for gas analysis. The conventional method for determining sample concentrations is to compare the integrated area of a spectral signal with that derived from a calibrated sample. The use of derivative detection dictated by frequency modulation would seem to rule out any such integration processes. Algorithms that relate the concentration to peak – peak (p – p) heights for rapid measurement, and to signal integrals over a set range for more precise determination, have been derived and tested however, and will be discussed in Section 4.1.

Measurement of samples at low concentration may be expected to yield signals badly contaminated by random and systematic noise. Techniques for noise reduction may be applied both in the design and the operation of the spectrometer and in post-detection data processing (Section 4.2). The utility and flexibility of these techniques have been greatly enhanced by the computing power now available from desk-top computers that can be attached on-line to a spectrometer in order simultaneously to process its output and optimise its working parameters. Some of the techniques that have been used are described in Section 4.3 and their development for measurements up to atmospheric pressure is discussed in Section 4.4.

1 Measuring the Spectral Line Profile and Area Using Frequency Modulation

MMW spectral linewidths are dominated by collisional pressure broadening though there is a fundamental underlying Doppler width given by Equation 1.30. For example, at 150 GHz this takes on the value 0.24 MHz for OCS at 290 K. Superposed on this is a pressure broadening whose magnitude is typically 60 kHz Pa^{-1} for samples in air-diluted mixtures. Thus up to $\sim$8 Pa the linewidth is sensibly constant whilst the profile has a Gaussian shape (Figure 1.4). Above this it increases linearly with pressure whilst the line follows a Lorentzian profile.

Once the sample pressure has become sufficiently high that Doppler broadening may be neglected, the peak signal from an unsaturated spectral line becomes independent of the pressure for a particular sample composition (Section 1.2). In this regime α_{max} depends on frequency, temperature, linewidth and the molecular concentration of the state under study. As an isotopomer, stereoisomer or vibrational excimer, the species concentration will be less than that of its parent.

By taking scans of an absorption profile over a sufficient frequency range, both the peak signal and the spectral area may be determined. This is normally desirable because both peak signal and linewidth for a sample of particular concentration vary somewhat with the nature of the diluent gases (Figure 4.1a), whereas the area does not (Figure 4.1b). Spectral peak area measurement

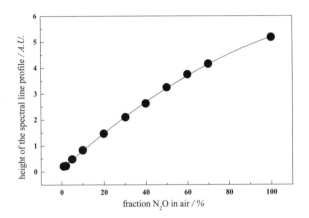

Figure 4.1a *Best fit line to measured line heights ●. The data are for nitrous oxide at 175926 MHz and 13 Pa*

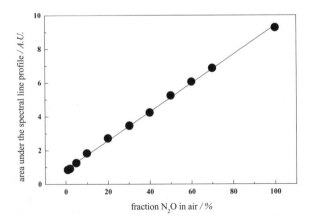

Figure 4.1b *Best fit line to measured line areas ●. The data are for nitrous oxide at 175926 MHz and 13 Pa*

therefore makes a more reliable calibration procedure. In the pressure broadened region the profile takes the form:

$$\Delta P/P = \frac{x\alpha_{max}L}{1 + \left(\dfrac{v - v_0 + \Delta v(t)}{w}\right)^2} \tag{4.1}$$

where the sample, present at fractional abundance x, absorbs power ΔP when the incident power P traverses a path L within it, v and v_0 represent the actual and peak absorption frequencies respectively, $\Delta v(t)$ is the instantaneous frequency shift caused by the modulation, and w is the HWHM.

The signal after detection is passed through a phase-coherent detector tuned to twice the modulation frequency ω. The signal at 2ω is the appropriate Fourier component of Equation 4.1 and has been derived for the case of sinusoidal modulation by Baker *et al.*[1] This signal is not of pure derivative form, but possesses a finite integral (Chapter 6).

A much simpler expression for the peak height is obtained, however, if a digital modulation of the form shown in Figure 4.2B is applied in which $\Delta v(t)$ jumps between peak positive and negative values $\pm \Delta v$ with an intermediate pause at zero. This is synchronised to waveform A and detected by waveform C. If the source is tuned to the peak frequency v_0 its modulated frequency simply jumps back and forth between the line peak and a symmetrical offset on each side, generating a square wave of angular frequency 2ω and amplitude:

$$\Delta P/P = x\alpha_{max}L\left[1 - \frac{1}{1 + (\Delta v/w)^2}\right]$$

$$= x\alpha_{max}L\left[\frac{(\Delta v/w)^2}{1 + (\Delta v/w)^2}\right] \tag{4.2}$$

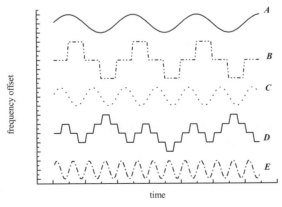

Figure 4.2 *Digital modulation strategies for phase-coherent detection. A is the synchronising waveform; B & C and D & E are modulation and detection pairs. For full explanation see text*

At low pressures where Doppler broadening dominates, w is constant and the absorption signal rises linearly with pressure. When pressure broadening takes over, the coefficient γ_{max} becomes pressure-independent and the spectral signal reaches a maximum, thereafter declining with increasing pressure as the applied FM fails to scan over the complete line width. The optimum is reached at a pressure somewhat greater than the 8 Pa quoted earlier. There is little to be gained by increasing the modulation depth beyond the value needed to attain this, a $p - p$ depth $\sim$240 kHz, because that would serve only to increase the size of the underlying frequency-dependent background without affecting the spectral signal.

For a single-pass FM spectrometer working at 150 GHz the best operating condition is to minimise background, by controlling power reflection within and at each end of a long-path absorption cell, and to apply a modulation depth of $\sim$240 kHz $p - p$ at a sample pressure $\sim$8–13 Pa. By contrast in a cavity spectrometer, the optimum modulation depth is governed by the cavity width, rather than the sample linewidth (Section 2.4).

2 Digital Modulation Techniques

It has been shown in Section 4.1 that replacing sinusoidal modulation by a digital form makes for a simpler expression for the spectral signal, and that the modulation amplitude can be selected to maximise its intensity. In fact the waveform demonstrated in Figure 4.2B, synchronised to waveform A, synthesises the second difference of the spectral function:

$$\delta^2 F(\nu) = 1/2 F(\nu) - 1/4 (F(\nu + \Delta\nu) + F(\nu - \Delta\nu)) \tag{4.3}$$

which serves also to remove linear background and to convert quadratic background to a constant pedestal. It is possible to carry background reduction even further by synthesising yet higher order differences; *e.g.* the fourth difference:

$$\delta^4 F(\nu) = 3/8 F(\nu) - 1/4 (F(\nu + \Delta\nu/2) + F(\nu - \Delta\nu/2))$$
$$+ 1/16 (F(\nu + \Delta\nu) + F(\nu - \Delta\nu)) \tag{4.4}$$

can be generated by the waveform shown in Figure 4.2D and detected with the waveform E. In principle this will remove all background variations up to the third order. The increased number of small steps required results, however, in a nett loss of signal at the peak frequency. Using a waveform of the same $p - p$ amplitude as, for example, Figure 4.2B produces a maximum signal of:

$$\Delta P/P = x\alpha_{max} L \left(3/8 - \frac{1/2}{1 + (\Delta\nu/2w)^2} + \frac{1/8}{1 + (\Delta\nu/w)^2} \right)$$
$$= x\alpha_{max} L \left(\frac{3/32 \ (\Delta\nu/w)^4}{(1 + (\Delta\nu/2w)^2)(1 + (\Delta\nu/w)^2)} \right) \tag{4.5}$$

some 7 times smaller than that generated by the second difference waveform when modulation depth and linewidths are equal. Although the fourth and other difference waveforms find use in some special cases when the highest spectral resolution is called for, we have found that the second difference waveform provides the best compromise for retaining signal amplitude whilst still reducing unwanted background.

Because the collision broadening parameter can be a marked function of the sample composition (Section 1.2), it is not usually sufficient to optimise and measure peak spectral signals in analytical spectrometry. Some measure of their width or area is called for if reproducible concentration measurements are to be achieved, The first- and second derivative-like signals characteristic of FM spectrometry whether analogue or digital, are particularly ill-suited to total area measurement and alternative width estimations need to be made. One relatively simple procedure is to seek the zero crossing points of the second difference signal. Taking Equation 4.3 as an example, these occur when:

$$2F(\nu) = F(\nu + \Delta\nu) + F(\nu - \Delta\nu)$$

or

$$\frac{2}{1 + ((\nu - \nu_0)/w)^2} = \frac{1}{1 + ((\nu - \nu_0 + \Delta\nu)/w)^2} + \frac{1}{1 + ((\nu - \nu_0 - \Delta\nu)/w)^2} \quad (4.6)$$

Equation 4.6 is satisfied when

$$(\nu - \nu_0)^2 = (w^2 + \Delta\nu^2)/3 \quad (4.7)$$

If we take this equation together with Equation 4.2 we can write:

$$\frac{w(\Delta P/P)_{max}}{x\gamma_{max}L} = \frac{w\Delta\nu^2}{w^2 + \Delta\nu^2} = \frac{\Delta\nu^2\sqrt{(3(\nu - \nu_0)^2 - \Delta\nu^2)}}{3(\nu - \nu_0)^2} \quad (4.8)$$

The product $w(\Delta P/P)_{max}$ is proportional to the area of the unmodulated line (Section 1.2), and so in Equation 4.8 we have a relation between this area, the inherent line strength, the sample concentration and a set of frequency parameters readily determined by experimental measurement from the frequency modulated profile. This technique depends importantly, however, on making a good estimation of the background signal, which can become quite difficult in the case of weaker spectra.

An alternative procedure that has been tested by us is that of measuring the area of the derivative signal between pre-set points such as its zero crossings,[2] that might be expected to yield comparable results. There were found no significant differences between relative concentrations evaluated from line peak and area measurements of oxygen in air mixtures. Results for nitrous oxide and air mixtures (Figure 4.1)[3] showed, however, a marked curvature of the peak measurements as a function of nitrous oxide concentration, whereas area measurements lay on a good straight line. This behaviour would be expected if the linewidth parameter of the pure sample differed markedly from that in air dilution.

It is not clear at the present time whether area integration or the application of Equation 4.8 will give the more reproducible results. The advantage of the former is that it takes account of broadening mechanisms other than those due to collisions alone. The disadvantage is that it takes longer to carry out a full spectral scan than does a rapid search for a peak and two zero crossing points. Both techniques suffer from the problem that any uncorrected background signal gives a spurious result.

3 Computerised Enhancement of Spectral Signals

Raw data extracted after a frequency scan of the spectrometer are obscured by noise as well as by any residual slope remaining after applying analogue or digital frequency modulation. Previously, the only way of dealing with the noise was to increase the time constant of the phase-coherent detector whilst scanning sufficiently slowly to avoid spectral distortion. Slope correction was achieved by carrying out comparative scans with the sample present and absent. Both procedures were acutely susceptible to false signals caused by slow drift or system glitches. Nowadays with the advent of powerful computation technology the recommended procedure is instead to collect data as rapidly as possible, reducing the effects of noise by averaging multiple scans and dealing with varying background by using computerised filtering algorithms.

In modern instrumentation spectral signals are collected and stored in digital form within appropriate channels of a microprocessor or a dedicated computer. The speed and power of commercial devices are now so great that these signals may be analysed in real time and the results of the analysis used to control the behaviour of the entire system. General computer algorithms suitable for these purposes are now widely available[4] but for optimal performance these need to be tailored to the properties and behaviour of particular instruments. In this section we consider specifically the collection and processing of data derived from FM MMW spectrometers.

The major ways in which FM MMW spectrometers show behaviour differing from the ideal are in the superposition of noise, background and drift on the desired spectral signals. Although noise may be readily removed by filtering and smoothing it must be realised that these procedures always distort the desired signal and in most cases produce a cosmetically enhanced rather than a fundamentally improved result. Recommended ways of dealing with noisy data include fitting them directly by a least-squares program to a parameterised model or taking a Fourier transform before applying an optimal Wiener filter.[4]

In the first case convergence may be expected to be slow, however, and the model parameters will be contaminated by large errors if the signal to noise ratio is poor. In the second case some prior assumptions about the profiles to be extracted must be made.

It is essential to obtain the largest signal to noise ratio from an instrument before selecting a filtering process that best matches the spectral feature sought. In quantitative spectrometry the position, profile and sometimes even the width of the desired features are known in advance. These contributions reduce the adverse effects of data smoothing and greatly aid the extraction of significant data even when obscured by noise.

The simplest smoothing procedure is a moving window average, in which each of $2N$ data points is replaced by the mean of itself and its nearest neighbours:

$$f_i = \sum_{n=-N}^{N} c_n y_{i+n} \qquad (4.9)$$

where $c_n = 1/(2N+1)$.

This averaging process reduces random fluctuations by a factor $(2N+1)^{\frac{1}{2}}$ whilst leaving any pedestal or linear background unaffected, though it does distort a quadratic background. More seriously, it broadens and decreases the height of any spectral feature present despite leaving its area unaffected.

A much more versatile smoothing technique is that by Savitzky and Golay.[5] This approximates the function between the range of points chosen by a polynomial, often a quadratic- or a quartic, and is capable of reproducing the shapes of both the spectral signal and its background whilst still reducing random fluctuations by the same factor as that quoted above. In particular, the slopes at the end of the fitted range merge continuously with those generated from adjacent data points, so that the smoothed data set shows no sudden jumps or kinks, but still faithfully reproduces the original spectral structure.

The presence of background caused by unabsorbed source power falling on the detector often presents a more serious problem than noise alone, because this generates a frequency-varying pedestal superposed on the signal sought by FM spectrometry. Procedures that involve integration over the profile are consequently compromised. It is possible in principle to collect and store such background data from blank runs for subsequent subtraction or data division. It may not always be convenient, however, to carry out a run in which the sample is absent, especially if the results of the scan are used to optimise and update spectrometer parameters. Furthermore such a procedure inevitably degrades spectrometer performance by introducing an additional source of noise through the subtraction or division process.

Instrumental drift, usually due to slow variations in the background signal, adds further to these problems. It introduces fluctuations into the subtraction and division processes described above and can vitiate attempts at comparison between signals taken at different times under apparently the same conditions. In particular, it makes extraction of weak signals by storing a background before starting a series of runs questionable, and throws doubt on those procedures which monitor spectral peaks alone without scanning over the profile to ascertain their background level.

If integrals over spectral profiles are to be determined experimentally, the most important parameter is the datum from which this profile is to be measured. If this level cannot be obtained directly for the reasons quoted above, it may be approximated by further numerical techniques. Again, the simplest is to carry out a moving window average over a range much wider than the spectral line itself. This will give a data set which correctly reproduces a linear background and in which the spectral line is much reduced in amplitude. Subtracting this from a Savitzky–Golay smoothed data set will yield a profile that is little different from that of the desired spectral line superposed on a zero background. Because the moving window average generates a broad feature whose integral corresponds to that of the original line, however, any further integration process must be confined strictly to the width of the smoothed feature and not range over the entire data set. Any more sophisticated procedure than this needs to involve some kind of background simulation that makes assumptions about its shape beneath the spectral line.

An alternative form of background and noise removal is to convolute the data with the expected form of the spectral signal. This may be done simply by replacing the coefficients c_n in Equation 4.9 with the amplitudes of the expected profile at an offset of n channels from its peak position, and carrying out this weighted smooth for each data point y_i. Although the expected lineshape is almost Lorentzian (Section 1.2), we have found it convenient to use a Gaussian profile for convolution purposes because of its lesser wing amplitudes, which permit truncation of the summation in Equation 4.9 at lower values of N.

The result of carrying through such a convolution is to enhance those components of the spectral signal which have a similar shape and width to those of the convoluting profile, and to reject noise components that do not. In fact, it may be shown analytically that the procedure converts a Gaussian profile of any width into a similarly positioned one whose width is the rms sum of those of the original and the convoluting function. Because its amplitude is reduced by the factor (1 + *original width/convoluting width*), only those features comparable to or narrower than the convoluting width are retained. It is interesting further to note that this entire procedure is mathematically equivalent to the recommended noise reduction process of taking a Fourier transform of the signal, applying a Gaussian filter of appropriate width, and then taking the inverse Fourier transform to obtain a noise-free signal.

Though a Gaussian convolution of this form is even more effective than Savitzky–Golay smoothing in improving the signal to noise ratio of spectral peaks, it has no effect on the background. A simple modification can be made to achieve this, however; instead of using a Gaussian profile for convolution it can be replaced with the first, second or even higher derivative of the same function.

The effect of this change is quite dramatic as witnessed between the unsmoothed and smoothed data in Figures 4.3 and 4.4 from measurements on OCS and HDO, both at low concentration. The transition in OCS is in a thermally populated vibrational state at 2.7 ppm abundance. The HDO was in natural water diluted with air and calculated from its vapour pressure to be at ~8 ppm concentration. Spectral signals are converted into the derivative form used, whilst more slowly varying background components are successively removed. A linear

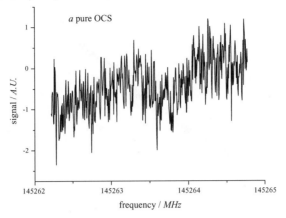

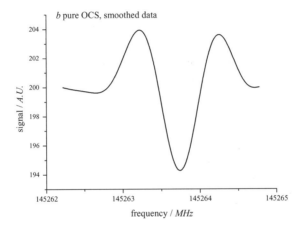

Figure 4.3   *Scans of the 12–11 transition of the 01⁻¹0 state of OCS at 145263 MHz and 2.7 ppm abundance:* (a) *before and* (b) *after Gaussian convolution and enhancement*

background is converted into a pedestal by a first derivative convolution and completely removed by a second derivative convolution, whilst a quadratic background becomes a pedestal in the second case. The outcome of the whole procedure is to generate sharp and narrow peaks superposed on a relatively flat background, an ideal combination for accurate signal measurement. Nearer to the limit of detection the difference between background and peak inevitably becomes less clear, exemplified in Figure 4.5 for oxygen. Knowing where the line is helps in the interpretation, but successive scans would reveal fluctuations in the amplitude of both line and background, as one would expect at the detection limit.

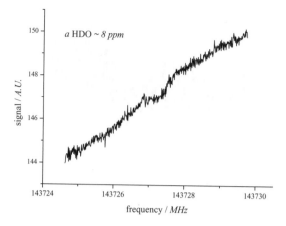

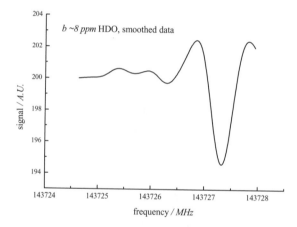

Figure 4.4   *Scans of the 4_{22}–4_{23} 143727 MHz transition in HDO at 20 °C and 13 Pa, concentration ~8 ppm in air:* (a) *before and* (b) *after Gaussian convolution and enhancement*

Unfortunately the process of background removal also removes any possibility of measuring concentrations from spectral integrals, because the signals are now true derivatives whose analytic integrals become identically zero. The relation-ships between the original spectral signal area and the heights and widths of digitally modulated signals derived in Section 4.2 still hold, however, leading to a convenient and rapid means of concentration determination from the signal height and its zero-crossing frequencies. Carrying out measurements in this way offers the potential of both simplicity and speed, because the scan range over which data need to be obtained can be several times less than that required for an integration.

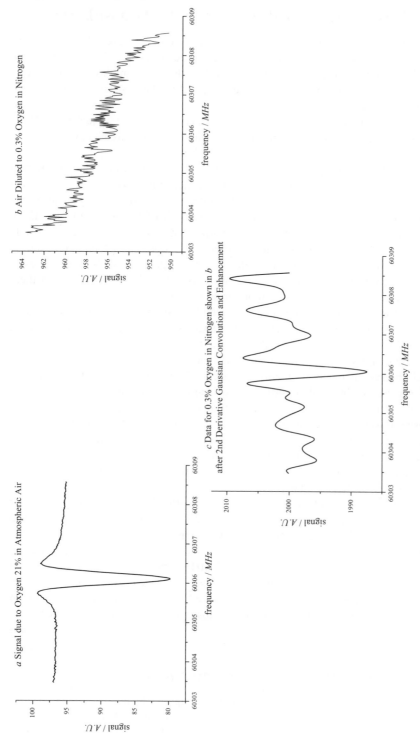

Figure 4.5   *Second derivative Gaussian smoothing and enhancement of an oxygen line. The signal in (c) is close to the limit of detection; the adjacent background structure was not reproducible, indicating that it was not due to spectral features*

4   Measurements at Pressures up to Atmospheric

Most published applications of microwave spectrometry have taken data from samples at Pa pressures. It is widely recognised, however, that the technique would be far more useful if it could deal directly with samples at atmospheric pressure and some recent work has been directed to this end and is discussed in Chapters 5 and 6. Linewidths at atmospheric pressure are ∼6 GHz making it essential for frequency scanning spectrometers to employ wideband MMW sources, but at the same time the requirement for source frequency stability is much reduced.

With direct spectral profile display it is necessary to determine the background level before a reliable lineshape fit can be carried out. This calls for frequency scans covering a range markedly greater than the ∼6 GHz linewidth and puts great demands on the power levelling as a function of frequency, of the MMW source and other components of the spectrometer. The fitted absorption profile must then be compared with the MMW absorption background, demanding careful system-gain calibration if these differ significantly. In precise work correction for changes in sample temperature will need also to be made, but there will be little demand for pressure monitoring or stability.

A derivative display, such as that produced by FM and subsequent convolution, can become markedly narrower than the spectral line from which it originates (Figure 3.9), and so is less difficult to track. The same algorithm when applied to the background signal removes both pedestals and linear slopes, so that the peak signal observed is much more closely attributable to the line itself. Determination of this peak height is often sufficient to characterise the absorption. The other side of the coin is, however, that loss of information regarding the background pedestal makes it difficult to determine the unabsorbed power level and hence the sample absorption coefficient. Furthermore, to apply a depth of frequency modulation comparable with the ∼6 GHz atmospheric pressure linewidth is out of the question for currently available MMW sources and the FM technique in its basic form is rarely used above 130 Pa.

One may, however, obtain the best of both worlds by using a cavity stabilised FM system.[6] Frequency modulating a high-Q cavity containing the sample and locked to the MMW source frequency will yield a signal corresponding to the cavity response damped by the sample. Only sufficient FM depth is required to cover the cavity bandwidth typically <1 MHz. The system background is an easily measurable signal from the cavity itself, when empty of any sample. In principle such a system can be stably tracked over an entire spectral width of 6 GHz or more and would thus allow the absorption signal from an atmospheric pressure sample to be measured.

5   Cavity Spacing Glitch Spectrometry

We have recently reported an interesting new approach to quantitative spectro-metric measurement.[7] As the Mark I semiconfocal Fabry–Perot cavity (Section 5.3) was tracked in synchronism with the MMW source through a sample

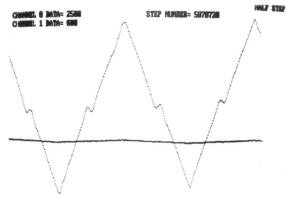

Figure 4.6 *Mirror spacing during successive down- and up-frequency scans of the locked cavity-servo system. The mirror spacing was monitored by a capacitance micrometer, the output being displayed on the ordinate axis vs. time, with 2 min sweep period. The glitch deviation was 2.5 μm, caused by an absorbing sample of natural isotopomer $O^{13}CS$ in OCS at 13 Pa*
(Reprinted from Rezgui *et al.*[7] with permission from Elsevier Science)

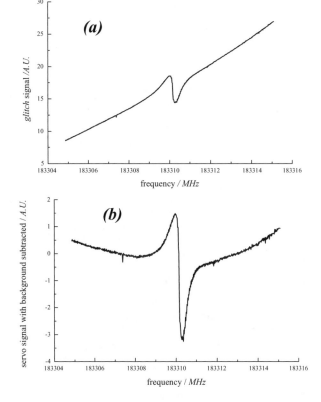

Figure 4.7 *Uncorrected* (a) *and background-corrected* (b) *glitch signal due to the H_2O $3_{03}-2_{12}$ transition*
(Reprinted from Rezgui *et al.*[7] with permission from Elsevier Science)

resonant frequency its spacing did not follow the source smoothly but showed a marked glitch as the spectral line was crossed (Figure 4.6).

This glitch is caused by interaction between the cavity and the sample resonance profiles. It may be converted into an apparent frequency shift of the cavity resonance that is directly related to the absolute absorption coefficient. The linewidth can be determined from the distance between the peaks of the glitch. This phenomenon became even more marked with the Mark II confocal Fabry–Perot cavity (Section 5.3), when it could be observed as a glitch in the correction voltage applied by the servo amplifier to the piezoelectric actuator of the moveable cavity mirror. Figure 4.7 shows a spectrum for the water line obtained in this manner.

The cause of the glitch proved to be inherent in the performance of the servo system when locking to a cavity profile whose Q was reduced by a narrower

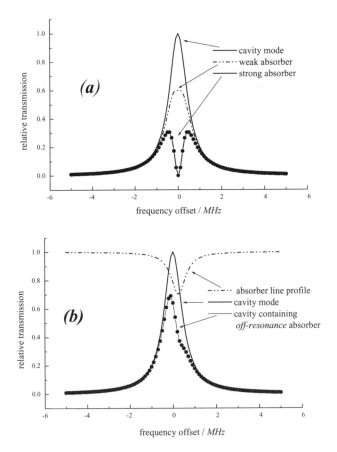

Figure 4.8 *Frequency offset response characteristics of the cavity Containing* (a) *a sample with the same resonance frequency, and* (b) *a different resonance frequency from that of the cavity*
(Reprinted from Rezgui *et al.*[7] with permission from Elsevier Science)

spectral line. The fall in Q and distortion arising from a spectral line approaching the cavity profile from one side, as occurs during a scan, are demonstrated in Figure 4.8. The phenomenon does have practical implications in that it is potentially as sensitive as direct absorption measurements using conventional Schottky barrier diode mixers although neither of those matches the sensitivity of a helium cooled bolometer. The ideas are still being developed for the exploitation of this effect and will be reported in the future.

References

1. J.G. Baker, N.D. Rezgui and J.F. Alder, Quantitative MMW Spectrometry (III) Theory of Spectral Detection and Quantitative Analysis in a MMW Confocal Fabry–Perot Cavity Spectrometer, *Anal. Chim. Acta*, 1996, **319**, 277–290.
2. J.G. Baker and J.F. Alder, Quantitative MMW Spectrometry (IV) Response Curves for Oxygen in Carbon Dioxide and Nitrogen at 60 GHz, *Anal. Chim. Acta,* 1998, **367**, 245–253.
3. N.D. Rezgui and J.G. Baker, unpublished work, 1996.
4. W.H. Press, B.P. Flannery, S.A. Teukolsky and W.T. Vetterling, *Numerical Recipes,* Cambridge University Press, 1986.
5. A. Savitsky and M. Golay, Smoothing and Filtering Functions for Analytical Data, *Anal. Chem.,* 1964, **36**, 1627–1632.
6. N.D. Rezgui, J. Allen, J.G. Baker and J.F. Alder, Quantitative MMW Spectrometry (I) Design and Implementation of a Tracked MMW Confocal Fabry–Perot Cavity Spectrometer for Gas Analysis, *Anal. Chim. Acta*, 1995, **311**, 99–108.
7. N.D. Rezgui, J.G. Baker and J.F. Alder, Quantitative MMW Spectrometry (V) Observation of Dispersive Gas Spectra with a MMW Confocal Fabry–Perot Cavity Spectrometer, *Anal. Chim. Acta,* 2001, **433**, 269–279.

CHAPTER 5

Cavity Spectrometer Designs and Applications

Although cavity spectrometers have been developed since the early days of microwave spectroscopy, Townes and Schawlow[1] refer to work from 1947, relatively few were ever developed for quantitative analytical work. Most early centimetre wavelength spectroscopy required many GHz bandwidth for the study of molecular structure and the compact design of cavity spectrometers never compensated in laboratory work for the narrow bandwidth they exhibited. The need for Stark modulation electrodes and the requirement for minimal line distortion over a wide spectral region directed the use of low-Q waveguide absorption cells. Wide-band high-Q cavities have been described.[1] The high Q is derived from the large volume to surface area ratio of the cavity; the volume imparts greater stored energy, the surface area greater resistive losses. They are large volume (tens $\times$ L) devices of little practical use in analytical applications.

1   Previous Commercial Microwave Spectrometers

Spectroscopic laboratory work was largely served by the Hewlett Packard spectrometers of the mid-1970s, some of which still appear in the literature even now. The quantitative spectrometer of Cambridge Instruments from the same era had a short commercial life, and aside from a few bespoke spectrometers for particular applications[2,3] there has not been a commercial off the shelf instrument for general analytical use since then.

2   Fabry–Perot and Waveguide Cavity Spectrometers

The most comprehensive description and review of cavity spectrometers in particular was by Varma and Hrubesh.[4] Referencing earlier work,[5,6] they used a design based on a semiconfocal configuration of the Fabry–Perot cavity. The source was a BWO and most unusually they used Zeeman effect magnetic field modulation. They were studying free radicals in the gas phase and were thus able to employ that method, with its added discrimination against the majority population of diamagnetic species. The cavity worked in the 18–40 GHz K-band and had a sensitivity $\sim 2 \times 10^{-9}$ m^{-1}. Lines of the species NF$_2$ with absorption coefficients between $10^{-7} - 10^{-8}$ m^{-1} were measured with a signal to noise ratio of $>5:1$.

They described a portable battery-powered semi-confocal Fabry–Perot cavity spectrometer. A Gunn device was critically coupled to, and thus its oscillation frequency dictated by the resonant cavity. It was designed for atmospheric analysis and achieved a limit of detection ~ 30 ppb formaldehyde in air, with a similar instrument detecting ~ 10 ppb ammonia in air.[4] They also alluded to rather simple designs of spectrometer for non-demanding analyses. A simple varactor tuned Gunn device coupled into a tuned waveguide cavity working presumably at the 22 GHz water line was described, with limit of detection for water claimed as 1 ppm. The water was introduced at low pressure by pumping against a membrane separator. The idea of a piezoelectric-tuned FM microwave absorption cavity spectrometer was patented by Leskovar *et al.*[7] The applicants described in the patent a semi-confocal Fabry–Perot resonator with both mechanical and piezo-electric tuning of the cavity.

The work of Thirup *et al.*[8] drew on many of the earlier developments in cavity spectrometers to produce a simple, compact analytical instrument suitable for routine gas mixture analysis. It was employed at frequencies up to 25 GHz with Stark field modulation of the analyte gas and incorporated an interesting approach to cavity spectrometry that would make it suitable for process control applications. The important feature was that the sinusoidal electric field impressed on the target analyte gas in the cavity gave rise to Stark modulation of the power reflected from the cavity. The Gunn device source was critically coupled to the cavity and its oscillation frequency dictated by the cavity complex impedance. Control of the cavity resonance was by a lightweight copper clad circuit board made into the plane mirror of the semiconfocal cavity. The plane mirror was maintained in position by the conical diaphragm of a loudspeaker onto which it was glued. The servo control circuit was tuned to the second harmonic of the Stark field modulation frequency which became zero when the cavity resonance moved the Gunn oscillator frequency to a known position very close to the peak absorption frequency. At that point the fourth harmonic signal was maximal and proportional to the fraction of target analyte gas in the cavity. The modulation harmonic signals were detected in a Schottky barrier diode mixer and passed to the second and the fourth harmonic phase-coherent detectors. For all its simplicity the spectrometer had a measured sensitivity of $\sim 2 \times 10^{-7}$ m^{-1} over the waveguide band 18–26 GHz. In principle it could operate at any frequency in that band, but in practice the tuning and coupling of the cavity to the Gunn device was not simple; it was best operated at a fixed frequency $\pm \sim 100$ MHz. In that mode it was successfully applied to the determination of ammonia, acrylonitrile and methanol in air, and water in air and liquefied petroleum gas (LPG) vapour. The water concentration working range was linear from 130 mg m^{-3} to >10 g m^{-3} and the presence of LPG vapour as the matrix had no discernible effect on sensitivity or linear range.

In their respective laboratories, Thirup *et al.* demonstrated that the instrument could work as a marginal oscillator spectrometer with the detected signal being transformer coupled out of the Gunn device bias circuit (Section 3.1). The spectrometer was applied to the determination of methanol and ammonia in air.[9] As a single analyte monitor the simplicity of the instrument showed promise, although design and optimisation of the microwave components was far from trivial. The attraction of FM rather than the high-voltage Stark modulation, higher frequencies and compact designs directed work down those paths instead (Chapter 6).

One of the most deeply developed analytical spectrometric methods in recent years was by Zhu *et al.*[10] They constructed a cavity and varactor tuned Gunn source Stark spectrometer using a 0.5 m waveguide transmission cell. The Gunn oscillator was frequency modulated *via* a varactor by a sawtooth wave of fundamental frequency 38 kHz which resulted in a signal filtered by the cavity to yield a frequency profile of the FM Gunn output approximately the same FWHM as the spectral line. This signal was passed through the wideband low-Q absorption cell where the absorption line was Stark modulated in a sinusoidal field. The interaction of the Stark amplitude modulation with the source FM gave rise to sidebands of the Stark modulation frequency 3.9 kHz superimposed on a carrier signal at twice the FM rate 76 kHz. The 76 kHz signal was synchronously detected to separate the amplitude modulation upper and lower sidebands that are mutually in antiphase. The sidebands were summed in separate phase sensitive detectors locked to 3.9 kHz. When the absorption centre frequency was coincident with the Gunn oscillator frequency, the sidebands summed to zero and the resulting error signal could be used in a conventional frequency control loop. The correction signal was applied to the thermostatic control of the Gunn oscillator cavity, thus altering its resonance frequency through its dimensions. The resulting spectrometer was stable and reproducible when operated at 23 GHz for the determination of ethylene oxide in air.

3   Double Resonance Spectrometers

Earlier work by Lee and White[11] was repeated by Fehse *et al.*[12] using a double resonance crossed Fabry–Perot spectrometer with frequency modulation of the probe source. Both worked in the 26–40 GHz region for the probe frequency, with 15 GHz and 23 GHz pump radiation respectively. The work was not reported in any great depth but served to illustrate that the technique was viable according to both sets of workers, if less sensitive than conventional Stark spectrometry.

The approach is interesting, because with double resonance techniques it is possible in principle to uniquely identify a component from a single measurement in even a complex gas mixture. This arises because the probability of two spectral lines each from two molecules in the same mixture overlapping is so vanishingly small.[13,14] Double resonance methods have therefore a certain charm for the analytical scientist who is always seeking methods showing absolute identification possibility. The methods as configured, *e.g.* Andrews *et al.*[15] and references therein, are, however, complex and expensive. They would be probably too complicated for routine use but could have applications for a specific target analyte in a complex matrix.

4 Recent Developments in MMW and Higher Frequency Spectrometers

Krupnov,[16] in a review of the then current state of *sub*-MMW spectroscopy in his home-institute, reflected on the absence of significant review papers in the preceding decade. He went on to describe the developments in high-frequency MMW spectrometers and particularly the development of laboratory versions of synthesisers up to 500 GHz and spectrometers up to 1.5 THz, based around their BWO designs.

More recent work on MMW spectrometers has indeed focused on BWO rather than Gunn sources, largely because of their wide operating bandwidth. Petkie *et al.*[17] described a fast scan sub-MMW spectroscopic technique (FASSST) based on a BWO. The source was characterised by a high spectral purity $\sim$10 kHz bandwidth carrier working from 240 to 375 GHz. The absorption cell was low-Q and the detector was an InSb bolometer operating at 1.5 K. The spectrometer was able to sweep 100 GHz in 1 s and smaller ranges *pro rata* with a resolution of 0.1 MHz using signal averaging. This remarkable achievement was due *they claim* to the exceptional spectral purity and tuning range of the BWO source, and could be well suited to monitoring unknown gas mixtures for different components. Although the sensitivity of the instrument was not quoted, strong lines of HNO_3 were reported as having a signal to noise ratio of $\sim 10^4 - 10^5$ in a 1 MHz bandwidth. The method clearly offers tremendous potential at these higher frequencies.

Another approach to MMW spectrometers is based on the *Orotron*[18] This device, called after the Russian words for an open resonator and a reflection grating, was a semiconfocal Fabry–Perot cavity (Figure 5.1) with the plane mirror having a reflection grating ruled upon it. The cavity, with $Q \sim 10^4$, produced a spectral bandwidth without frequency locking $\sim$10–15 kHz and output power was 3–10 mW over 90–150 GHz.

The device is unique in that the absorption cavity is part of the MMW generation structure. It is a type of non-relativistic free electron laser and the resonant frequency of the Fabry–Perot structure gives rise to the laser frequencies that depend upon the 400–1000 V accelerating voltage applied to the plane grating. A magnetic field 0.6 T around the cavity directs the electron beam. The cavity was separated into the low-pressure acceleration region and the intermediate pressure region where the analyte gas could be introduced.

Detection of absorption was achieved by source FM and monitoring changes in the current when the Orotron was in a state of marginal oscillation using the usual phase-coherent detectors. The device under those conditions operated as a square law detector with sensitivity $\sim$1 AW^{-1}. The resulting sensitivity of absorption was $\sim$3–5 $\times$ 10^{-8} m^{-1} with a 1 Hz bandwidth receiver and they report analysis of CO, CH_3OH, CH_3COCH_3 and OCS in admixture with water or air.

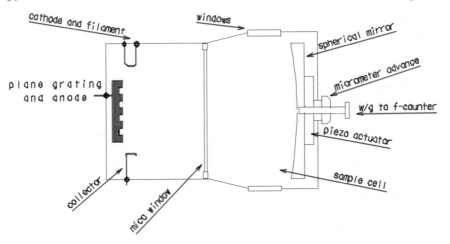

Figure 5.1 *Schematic diagram of an Orotron. The grating has 100 lines spaced much less than one wavelength. The surrounding magnet is not shown. Dimensions are 200 × 100 × 100 mm. Working pressure is typically 13 Pa*
(Adapted from Dumesh et al.[18] with permission from Elsevier Science)

Surin[19] used the Orotron oscillator as a tuneable source of coherent MMW radiation to study the spectroscopy of SiH_4 and ND_3 in the 90–160 GHz range. The gas was introduced into a cell placed in a Fabry–Perot cavity $Q \sim 10^4$. The absorption signals were detected from variation of the Orotron electron beam current. No phase or frequency lock schemes were necessary and resolution achieved was sufficient to resolve the lines' Doppler profiles; sensitivity was estimated as $3–5 \times 10^{-8}$ m^{-1}.

The Orotron is a remarkable device, comprising no MMW semiconductors or cryogenically cooled bolometer, but possessing nonetheless better sensitivity than most cavity spectrometers. The penalties are the mechanical complexity, a magnet, and importantly, risk of total system failure if the interior became corroded or contaminated by the target analyte. The last potential problem is not unique to an Orotron, however: other designs also risk that catastrophe!

Vaks et al.[20] have designed a novel spectrometric approach for gas analysis based on a BWO and low-Q absorption cell. The BWO spectral output was split between a reference cell and a 1 m waveguide measurement cell. The gas in the 0.2 m pathlength waveguide reference cell was used to lock the BWO frequency to the spectral line at the 304 GHz OCS spectral line and the frequency modulated spectral source directed to the analytical cell. Detection was with a Schottky barrier diode mixer.

The unusual feature of their approach was in the frequency stabilisation, signal recovery and background discrimination, which are problems inherent in all spectrometer designs. The BWO frequency was stabilised using a reference cell containing a sample of the gas under test. Its spectral line was used as a frequency discriminator (Section 3.3). The important feature was its resolution of phase information of the detector modulation signal with respect to the input modulation signal.

The authors noted that most workers ensure the modulation frequency f_m of a spectral source is much less than the FWHM of the spectral line: $\Delta\nu \gg f_m$. This is to minimise effects due to broadening and distortion of the recovered line profile information. The FWHM of the line is however due to a spectral process (Section 1.2). They remarked that if the modulation period $1/f_m$ was of the same order as the relaxation time of the molecule from its excited state then there was a significant phase shift of the recovered signal compared with the modulation signal. Critically important, however, is that the signal due to the background absorption, which is not due to spectral processes and is a broad bandwidth response, remains practically in phase with the modulation frequency. By measuring the recovered signal in quadrature to the modulation signal therefore, the influence of the background could be largely reduced. They demonstrated this effect using ratios $f_m/\Delta\nu \sim 0.1$, 1 and 2, with the best results happening at 2, close to the theoretical optimum of 2.2 showing some signal to noise enhancement and removal of the background slope.

The principles espoused in the work of Vaks *et al.*[20] are applicable to any spectral source and transition. There would, however, be a problem in applying it directly to a high-Q cavity spectrometer in that the cavity resonance shape and bandwidth are similar to those of the line, and background absorption would not be corrected. One might overcome that problem by use of a dual-phase lock-in amplifier with the cavity held at resonance as the spectral line was swept. The cavity tracking could be achieved by mounting a lightweight Fabry–Perot cavity mirror on a piezoelectric positioning device, locked to the in-phase component of the recovered signal.

It is questionable whether this approach would be any better than the post-detection data manipulation techniques described in Section 4.3. The trade-off is between complexity of instrumental design and operation, programming complexity and post-measurement data processing time. As long as the instrument response could be correctly and precisely modelled in software, the balance would probably lie on the post-analysis processing side.

5 Atmospheric Pressure MMW Spectrometers

One of the more significant papers of recent times in quantitative MMW spectrometry is by Krupnov *et al.*[21] They have addressed the measurement of broad spectral lines at atmospheric pressure using a Fabry–Perot cavity. The measurement of the full spectrum of even one line typically 6–8 GHz FWHM at atmospheric pressure has proved problematic and, apart from the extensive work of Liebe[22] on oxygen, relatively little studied.

The key to their success has been a broad bandwidth BWO phase locked to an 8–12 GHz synthesiser that defines the BWO centre frequency. The locked source was frequency modulated by a second 20–40 MHz synthesised sweeper. The Fabry–Perot cavity comprised a pair of 120 mm diameter $\times$ 240 mm radius of curvature silver-plated mirrors. Their spacing could be adjusted between 250 and 420 mm. The frequency sweep was achieved by stepping the frequency successively between the n longitudinal modes. The condition for resonance is that the mirror spacing $d = n\lambda/2$. For a given value of d, the cavity is therefore resonant at $\lambda_n = 2d/n$, corresponding to about 0.5 GHz change between two consecutive modes. By stepping and sweeping at each resonance, the authors were able to recover a full waveguide band spectrum in about 1 s. They demonstrated measurements for oxygen at 60 and 118 GHz, and water at 183 GHz with spectrometer sensitivity $\sim 4 \times 10^{-7}$ m^{-1}.

Significantly they showed that the spectral profile of the water line could be accurately described by the van Vleck–Weisskopf model (Equation 1.36), indicating guidelines for the fitting of spectral profiles to other absorbers in a mixture. This would permit their deconvolution from the rest of the spectrum, spectral profile area measurement and hence analyte quantification.

The key to their success was the high-Q ($\sim 6 \times 10^5$) cavity and precisely controlled broad band BWO source. The rf sweep synthesiser had a switching time of ~ 200 ns and time between switching of 58 μs, permitting precise and fast data acquisition at each of the 100–200 mode steps possible. Detection was achieved using a Schottky barrier mixer diode, with the usual phase-locked signal recovery electronics and digital data processing.

Solid state sources are not able presently to achieve the required power output over such a wide band; 10% of the centre frequency would represent a typically good -3 dB power bandwidth. For single component measurements it is not necessary to scan the entire waveguide band, the 6–8 GHz FWHM of typical lines at 50–200 GHz, could be adequately covered by a Gunn device. This would be an attractive method for dedicated measurements at atmospheric pressure of flame combustion products in smoke stack and engine exhaust effluents, e.g. oxygen and carbon monoxide whose concentrations at >1 ppm are particularly important.

The work of Rouleau et al.[23] is a recent insight to atmospheric pressure measurements on gas and liquid phases by GHz frequency dispersion methods. Although the work was developmental in character it does serve to illustrate the techniques available using MMW technology on higher pressure condensed phase systems. In this study they used a pair of cylindrical cavities working in the TE$_{112}$ mode configured in a balanced bridge configuration driven from a 3.5 GHz signal generator. Using source amplitude modulation and lock-in amplifier detection they achieved low-ppm sensitivity for water in Ar, He, H$_2$, CH$_4$ and SF$_6$. They also described the response–concentration relationship using the Clausius–Mossotti theory for summation of molar polarisability, that is the basis of these dielectric loss measurements and also applicable to the MMW spectrometry of liquid mixtures. The dielectric spectroscopy of liquids is discussed in detail in a recent text by McHale[24] and has previously been applied successfully to water determination in oils in our own work.[25]

The trade-off between the simplicity of atmosphere pressure work, spectral resolution, method selectivity and sensitivity strongly favours it for process analysis compared with even the modest vacuum pumping requirements of MMW spectrometry at the usual pressures of tens of Pa. Where the balance occurs will of course be application dependent. The scope for ingenuity in the design of self-contained gas sampling systems with pressure reduction to intermediate values between 10 Pa and 1 atm is clear. The optimum is unlikely to be at atmospheric pressure for any other than the simplest mixtures containing particularly oxygen, because other methods, *e.g.* laser diode infrared spectrophotometry, hold the attraction of cheapness and simplicity. The greatest potential for MMW spectrometry lies still in the low-pressure analysis of many-component mixtures without prior chemical separation being required. Isotopic and positional isomers, stereoisomers, free radicals and molecules in excited states can all be determined by low-pressure MMW spectrometry. Not only is their possible analysis attractive, but the use of naturally abundant isotopes and vibrational level population characteristics as internal calibration parameters is another feature of this versatile technique that has yet to be fully exploited by analytical scientists.

References

1. C.G. Townes and A.L. Schawlow, *Microwave Spectroscopy*, Dover Publications, New York, 1975.
2. J.F. Alder, Modern Microwave Technology Revitalises a Specific and Sensitive Analytical Technique, *Phil. Trans. Roy. Soc. London*, 1990, **333**, 19–27.
3. J.F. Alder, M.F. Brennan, I.M. Clegg, P.K.P. Drew and G. Thirup, Application of Microwave Spectrometry, Permittivity and Loss Measurements to Chemical Analysis, *Trans. Inst. Meas. Control*, 1983, **5**, 99–111.
4. R. Varma and L.W. Hrubesh, Chemical Analysis by Microwave Rotational Spectroscopy, in *Chemical Analysis*, ed. P.J. Elving, J.D. Winefordner and I.M. Kolthoff, Vol. 52, John Wiley and Sons, 1979.
5. L. Hrubesh, R. Anderson and E. Rinehart, *Rev. Sci. Instr.*, 1971, **42**, 789.
6. L. Hrubesh, *Radio Sci.*, 1973, **8**, 167.
7. B. Leskovar, H.T. Buscher and W.F. Kolbe, US Patent 4 110, 686, August 29, 1978.
8. G. Thirup, F. Benmakroha, A. Leontakianakos and J.F. Alder, Analytical Microwave Spectrometer Employing a Gunn Diode Locked to the Rotational Absorption Line, *J. Phys. E: Sci. Instrum.*, 1986, **19**, 823–830.
9. F. Benmakroha, Design and Construction of a Microwave Spectrometer, PhD Thesis, University of Manchester, 1986.
10. Z. Zhu, I.P. Matthews and A.H. Samuel, Design and Implementation of a Cavity Microwave Spectrometer for Ethylene Oxide Monitoring, *Rev. Sci. Instrum.*, 1995, **66**, 4817; *ibid.*, 1996, **67**, 2496; Z. Zhu, I.P. Matthews and G.W. Dickinson, *ibid.*, 1997, **68**, 2883.
11. M.C. Lee and W.F. White, A Cavity Absorption Cell for Double Resonance Microwave Spectroscopy, *Rev. Sci. Instrum.*, 1972, **43**, 638–640.
12. W. Fehse, D. Christen and W. Zeil, Microcomputer Controlled Microwave-Microwave Double Resonance Spectrometer Incorporating Two Crossed Fabry–Perot Resonators, *J. Mol. Struct.*, 1983, **97**, 263–270.
13. C.E. Jones and E.T. Beers, The Number of Frequency Measurements Necessary for the Microwave Identification of a Gas in a Mixture, *Anal. Chem.*, 1971, **43**, 656–659.
14. B. Vogelsanger, M. Andrist and A. Bander, Two-Dimensional Correlation Experiments in Microwave Fourier Transform Spectroscopy, *Chem. Phys. Lett.*, 1988, **144**, 180–186.

15. D.A. Andrews, J.G. Baker, B.G. Blundell and G.C. Petty, Spectroscopic Applications of Three-Level Microwave Double Resonance, *J. Mol. Struct.*, 1983, **97**, 271–283; D.A. Andrews and J.G. Baker, Pulsed Microwave-Microwave Double Resonance in Gases, *J. Phys. B*, 1987, **20**, 5675–5704; D.A. Andrews, N.J. Bowring and J.G. Baker, Novel Coherence Effects in Microwave-Microwave Double Resonance Pulse FT Spectroscopy, *J. Phys. B*, 1992, **25**, 667–678.

16. F. Krupnov, Present State of sub-MMW Spectroscopy at the Nizhnii Novgorod Laboratory, *Spectrochim. Acta, Part A*, 1996, **52**, 967–993.

17. D.T. Petkie, T.M. Goyette, R.P.A. Bettens, S.P. Belov, S. Albert, P. Helminger and F.C. DeLucia, A Fast Scan Submillimeter Spectroscopic Technique, *Rev. Sci. Instrum.*, 1997, **68**, 1675–1683.

18. B.S. Dumesh, V.P. Kostromin F.S. Rusin and L.A. Surin, Highly Sensitive MMW Spectrometry Based on an Orotron, *Meas. Sci. Technol.*, 1992, **3**, 873–878; B.S. Dumesh, V.D. Gorbatenkov, V.G. Koloshnikov, V.A. Panfilov and L.A. Surin, Application of Highly Sensitive MMW Cavity Spectrometer Based on an Orotron for Gas Analysis, *Spectrochim. Acta, Part A*, 1997, **55**, 835–844.

19. L.A. Surin, Intracavity MMW Spectroscopy of Molecules in Excited Vibrational States, *Vib. Spectrosc.*, 2000, **24**, 147–155.

20. V.L. Vaks, V.V. Kodos and E.V. Spivak, A non-Stationary Microwave Spectrometer, *Rev. Sci. Instrum.*, 1999, **70**, 3447–3453.

21. A.F. Krupnov, M.Y. Tretyakov, V.V. Parshin, V.N. Shanin and S.E. Myasnikova, Modern MMW Resonator Spectroscopy of Broad Lines, *J. Mol. Spectrosc.*, 2000, **202**, 107–115.

22. H.J. Liebe, Updated Model for MMW Propagation in Moist Air: Data on Oxygen and Water Absorption in the Atmosphere, *Radio Sci.*, 1985, **20**, 1069–1089.

23. J.F. Rouleau, J. Goyette, T.K. Bose and M.F. Frechette, Investigation of a Microwave Differential Cavity Resonator for the Measurement of Humidity in Gases, *Rev. Sci. Instrum.*, 1999, **70**, 3590–3594.

24. J.L. McHale, *Molecular Spectroscopy*, Prentice Hall, New Jersey, 1999.

25. J.F. Alder, I.M. Clegg and P.K.P. Drew, Microwave Determination of Water in Crude Oil Mixtures at K-Band (I), *Analyst*, 1984, **109**, 769–773; Microwave Determination of Water in Crude Oil Mixtures at K-Band (II), *ibid.*, 1986, **111**, 781–783.

CHAPTER 6

A Practical Frequency Modulated Spectrometer and Its Application to Quantitative Analysis

There are no laboratory or process analysis MMW spectrometers or kits commercially available at present to the authors' knowledge. To start designing *ab initio* an instrument for analytical method development is not easy, particularly if the reasons why particular choices of components were made are not clear. To alleviate this problem the authors would like to share their experiences in developing their own equipment with a few behind the scenes tips and anecdotes. This chapter therefore describes the design and evolution of an operating analytical spectrometer constructed by the authors with the help of their colleagues and students. Its purpose is to guide the reader through the steps and decisions taken that led to the system performance described in the papers of Baker, Alder *et al.*[1-5] Many of those were compromises influenced by financial as well as technical considerations. They may not be the same compromises that you *the reader* need to make and there may also be better solutions.

1 To Set the Scene

The goal of this programme was a bench-top instrument that could eventually be deployed in the field or on a plant, for routine automatic analysis of a gaseous sample stream for a few components at the 1 ppm to 100% concentration level. At the start of the project an overview was taken of the commercial spectrometer designs that had appeared on the market during previous years (Chapter 5). Their common feature was the use of microwave sweepers based on frequency stabilised BWO sources, gas samples maintained at Pa pressures and modulated by high electric fields, calibration by bridge spectrometry, and operation in the 10–40 GHz band. All these features led to cumbersome and expensive systems that were versatile but required specialised skills to operate them.

Our aim instead was to devise an inexpensive and transportable system that could easily be assembled in the laboratory, and that might be dedicated to the quantitative study of a few selected analyte species rather than the wide variety open to MMW spectrometry as a whole. In order to retain precision in measurement of sample signals, it would be necessary to deploy on-line computer control of the system as well as computer analysis of data. Allied to this it would be desirable to use a stand-alone mode of operation in which the measurements were made as insensitive as possible to design quirks, and to external environmental features such as temperature and pressure.

2 Signal Sources and Operating Frequency Band

The need for transportability coupled with operator- and intrinsic safety considerations dictates the use of solid state sources operating at low voltage. Of these, Gunn oscillators are the least susceptible to noise and, if locked to a high stability source, are easy to stabilise and to scan in frequency. At lower frequencies they are both inexpensive and widely available as components of intruder alarm systems, but their cost rises rapidly with increasing frequency, and becomes uncomfortable beyond about 100 GHz.

At higher frequencies the preferred option is to use a high-power Gunn source capable of delivering 50–100 mW in the 50–75 GHz band, followed by a doubler or tripler of 10–20% efficiency. Such a combination may be mechanically tuned over a 5 GHz band at the fundamental frequency, giving reasonable performance over 15 GHz at the third harmonic. At the same time the Gunn device is available as a source in its own right for spectrometry or calibration purposes, working at the fundamental frequency.

Although the analytical spectroscopist might well prefer a source tuneable through the entire MMW band of 30–300 GHz, as quoted for some BWOs, we are not aware of the existence of any single solid state device that currently matches this performance. Wide band sweepers available from some manufacturers often comprise a number of modules operating over smaller adjacent bands under computer control. The cost of these is for most of us prohibitive, but, more to the point, unnecessary. The attractive feature of MMW spectra is their even distribution throughout a wide spectral region. It is therefore possible to work in a restricted spectral range and yet still have access to all the species one may want to measure.

The best operating frequency will clearly be that at which the sample absorption is the greatest. Whilst the formula of Equation 1.48 suggests that this absorption increases indefinitely with frequency, no allowance has been made for the ever-decreasing thermal population of the states contributing to the transition. When this is factored in, one obtains the formula:

$$\alpha_{\max} = \frac{4\pi^2 N f_v i \mu^2 v_0^3}{3c(kT)^2 \Delta v} \{[\exp(-(v^2 - 2Bv)/4BkT)]/4\pi\varepsilon_0\} \qquad (6.1)$$

This formula is plotted in Figure 6.1 for OCS, whose properties are representative of a typical analyte.

The peak absorption coefficient of OCS, $\sim 10\ \text{m}^{-1}$, occurs at 462 GHz. This is by no means, however, the optimum working frequency due to the non-ideal behaviour of most MMW detectors. Commercial Schottky barrier mixer diode detectors show a quadratic roll-off in sensitivity at frequencies >100 GHz. If this is factored into Equation 6.1, the peak sample sensitivity occurs around 300 GHz, and the response is so flat that even at 100 GHz it has only fallen off by a factor of two. What is common to both curves is the dramatic fall-off in sample sensitivity at frequencies <100 GHz, reinforcing the point that the band 26–40 GHz is ill suited to high-sensitivity analytical spectroscopy.

Our final selection of a source was based on the consideration that the two common MMW active atmospheric constituents, oxygen and water, have spectral lines of sufficient sensitivity only around 60 GHz, at 118 GHz and at 183 GHz. The capability to measure them was a *must have* feature of our spectrometer; most other target species had spectral lines spread over a broader spectrum. With these considerations in mind a Gunn source was purchased to cover the 58–63 GHz band, together with a doubler and a tripler. In order to fill in the frequency gap between the harmonics a further source, comprising a 70–80 GHz Gunn plus a doubler was also used. Together these have provided access to the spectra of most species of interest

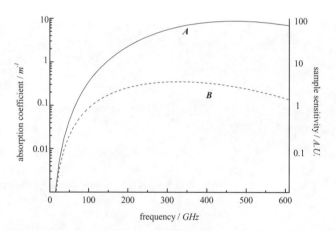

Figure 6.1 *Peak absorption coefficient A and sample sensitivity B for OCS Plotted vs. frequency. $B = 0.202\ cm^{-1}$; at 300 K: $kT = 200\ cm^{-1}$*

3   The Sample Cell and Sample Handling Techniques

In order to provide a detection limit of ~ 1 ppm for a sample of OCS, $\alpha_{max} = 1\ \text{m}^{-1}$ at 150 GHz, one needs to observe a power variation of ~ 1 ppm m^{-1} path length. Taking a typical commercial MMW detector with quoted tangential sensitivity: -40 dBm or 10^{-7} W, in 40 Hz bandwidth at 1 kHz modulation rate, it is clear that at an incident MMW power of $5-10$ mW one cannot expect to see absorption $<10-20$ ppm. This calls for a minimum path length of $10-20$ m. Such lengths are achievable in the laboratory only with some kind of folded optical structure, the most popular being (not surprisingly by now) the confocal- or semi-confocal Fabry–Perot cavity. To meet the above specification, it must be designed to have a loaded-Q $\sim$60 000, corresponding to an unloaded-Q $>$120 000, requiring a mirror reflectivity $>$0.995.

Our first attempt to construct a suitable cavity (Mark I) was an open structure based on a design by Thirup,[6] using two 100 mm diameter $\times$100 mm radius of curvature spherical brass mirrors (Figure 6.2). The rationale for the dimensions of those mirrors is given in Rezgui *et al.*[1] Briefly, there are several criteria that have to be met. First, a stability criterion that incorporates the mirror spacing L and the radius of curvature of the mirrors. Second, the Fresnel Criterion relating the mirror diameter D to the spacing and wavelength to ensure no diffraction losses $D^2/4L\lambda \geqslant 1$; see Section 2.1. Coupled with this is an estimate of the maximum volume required for the cavity. That is going to dictate the rate at which sample can be moved in and out of the cavity by the pump, and hence the response time of the instrument. There are other criteria too; for example, availability of PTFE rod of the required diameter was the key factor that dictated the maximum dimensions of the sample cup in our original design. The overall size of the

Figure 6.2   *The Mark I cavity spectrometer. The Perspex ring can be seen clamping the PTFE end wall to the sample cup and the thin section of wall in the centre is visible. The second mirror is seen from behind, mounted on its translation stage with a stepper motor and piezoelectric actuator*

instrument too is a factor to take into account: the bigger the cavity the heavier and more cumbersome it will become.

The sample was confined within a hollow *PTFE* cylindrical cup with one mirror serving as its cap. The gaseous sample was pumped into the enclosure made by the cavity and mirror *via* a needle valve restrictor, through an inlet port 6 mm in diameter near the periphery of the mirror. A diametrically opposite port was connected to the vacuum pumps. The cup base was a flat disc, 5 mm thick, halfway between the mirrors in the region of the beam waist. We soon discovered that this disc caused considerable dielectric loss, and therefore reduced its thickness to 1 mm over an area $\sim$30 mm diameter that gave a big improvement, and simultaneously formed a spatial filter which helped suppress non-axial modes; see Figure 2.5.

The mirrors were polished brass, and on one prototype polished stainless steel. We could have had them gold plated, which would have helped their chemical stability and reduced resistive losses. All our published results, however, are without any plating on the surfaces. At the time we were unsure of the precision to which the mirrors would need to be lined up, and provided one of them with three independent vernier translational drives. In the event the stability of the design was so great that these proved superfluous, and they have been discarded in subsequent systems.

The only problem that presented itself was one of microphony of the cavity resonance frequency, possibly due to fluctuations in gas pressure causing changes in the refractive index of the sample between the mirrors. With the incorporation of source FM, and piezoelectric actuator stabilisation and scanning of the mirror spacing, both described in detail in Section 4.1, the system was used to obtain signals from various analyte gases. Due, however, to the remaining dielectric loss due to the PTFE disc, we were unable to achieve a cavity-Q much greater than 3000. Limits of detection were generally worse than 100 ppm due mainly to the short optical path with this low Q. Nonetheless, it proved a simple, transportable and versatile system capable of carrying out routine measurements reproducibly on a regular basis.

The Mark II design shown in Figure 6.3 was designed to compensate for the faults discovered with Mark I. Diffraction losses were reduced by increasing the mirror dimensions to 200 mm diameter $\times$ 200 mm radius of curvature. The mirror position was adjustable around a spacing $\sim$200 mm. Dielectric loss was minimised by removing material from within the cavity and by enclosing the entire system inside an evacuated casing, which included a viewing port made of 12 mm thick Perspex. Sample and pumping ports were in the casing walls; see Figure 2.2.

The spectrometer was set on a simple vibration-free pressurised-air mounting and the microphony was effectively silenced. The mounting was made from a concrete paving slab, mass $\sim$20 kg, trimmed with lead bricks to be horizontal, resting on an inflated motorcycle tyre inner-tube; the spectrometer table legs were set in sand-filled pots. These simple provisions overcame successfully both the high-frequency shocks of general operation and the low-frequency vibration of our high-rise building.

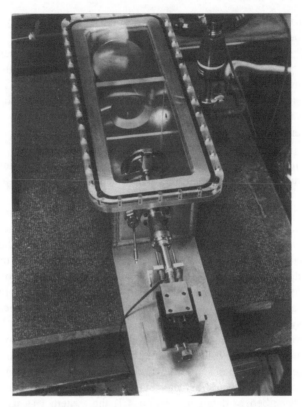

Figure 6.3   *The Mark II cavity spectrometer viewed through its Perspex cover. The spatial filter is visible in the centre of the case. The translation stage and piezoelectric actuator are seen in the foreground and operate via the bellows to move the near mirror. The coupling iris is in the centre of the far mirror. The waveguide couplings holding the mica vacuum windows can be seen either side of the case in the background. Sample inlet is on the left wall of the case not visible and outlet is on the right via the pressure sensor*

One mirror of the cavity was mounted rigidly on the baseplate of the casing, whilst the second was attached firmly to a brass rod and sleeve at the centre of which was a rod driven by a piezoelectric actuator capable of translations up to 100 μm. The rod/actuator assembly itself could be mechanically advanced by several cm, giving access to a large number of longitudinal cavity resonances, whilst the piezoelectric actuator was sufficient to traverse the resonant frequency through many linewidths. An additional feature of the Mark II system was the insertion between the mirrors of the spatial filter made of a Perspex disc (Figure 2.2) with a central 30 mm diameter hole corresponding to the waist dimension of the TE_{00n} modes. This proved effective in eliminating other unwanted modes of the cavity and in simplifying the response of the system as its excitation frequency was varied (Figure 2.5). Because the Perspex was outside the main region occupied by the electric field of the TM_{00n} mode, it had an insignificant effect on the Q of the cavity.

Coupling of the spectral source to this cavity was by transmission between two RG/99/U waveguide ports of i.d. 2.3×1.15 mm mounted in a vacuum-sealed metal cylinder that formed part of the backing to one mirror (Figure 2.6). To optimise the coupling the size of the coupling hole was adjusted by covering these ports with thin (100 μm) aluminium foil and piercing holes of the appropriate size and position in them. In most applications we have found it quite practical to dispense with this foil altogether, and just have the waveguide ends abutting the ~4 mm hole in the mirror. Whereas this is satisfactory for narrow bandwidth operation, for wideband work scanning several GHz much more careful matching of the cavity to waveguide would be required. In use this system displayed a $Q \sim 5 \times 10^4 - 2 \times 10^5$ in the 140–183 GHz band. A cavity resonance profile is shown in Figure 6.4. This spectrometer has been operational for some years and was used successfully to investigate many weak spectra, two of which for HDO and oxygen are shown in Figures 4.3 and 4.5.

In Figures 6.5–6.8 are shown spectra obtained for two common atmospheric gases. Water (Figure 6.5) is an important target for many analytical measurements. Its sensitivity by MMW spectrometry can most readily be judged by measuring the isotope HDO (Figure 6.6) in its natural abundance of 298 ppm in water. Figure 4.4 shows a scan of HDO in air at 8 ppm, near its limit of detection. Formaldehyde, a pernicious and lachrymatory component of exhaust and flue gases, is present in air samples at low concentration. Using the solid polymeric material paraldehyde as a convenient source of formaldehyde, Figure 6.7 shows a spectrum of $H^{13}CHO$ and Figure 6.8 the less abundant $HCH^{18}O$. The ultimate limit of detection we obtained for formaldehyde was as the $HCH^{17}O$ isotopomer at ~0.5 ppm.[3]

Figure 6.9 shows a transition in ethylene oxide, a gas of considerable interest

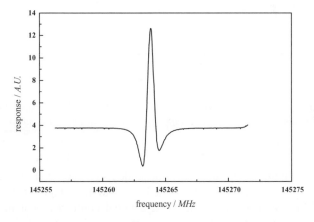

Figure 6.4 *Scan of the cavity profile showing* $Q = 200\,000$; *this corresponds to an effective pathlength of 60 m*

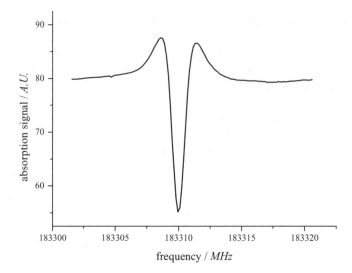

Figure 6.5   *The $3_{13}-2_{20}$ 183310 MHz transition of water at 13 Pa*

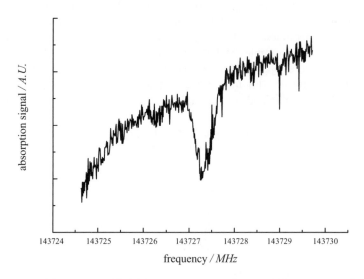

Figure 6.6   *The $4_{22}-4_{23}$ 143727 MHz transition in HDO at 13 Pa, in natural abundance (298 ppm) sampled from the vapour over water at ambient temperature*

industrially, and Figure 6.10 a doublet in acetaldehyde. The latter, along with the transition in the *trans*-rotamer of ethanol at 122551 MHz in Figure 6.11 illustrates the ability of MMW spectrometry to yield detailed conformational information. This is also shown elegantly with the spectra of the conformational isomers *endo*- and *exo*-2-norbornanecarbonitrile in the text of Varma and Hrubesh,[7] the two

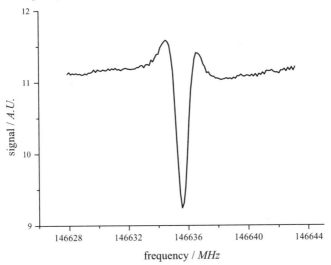

Figure 6.7 *The $2_{02}-1_{01}$ 146636 MHz transition in $H^{13}CHO$ at 26 Pa in natural abundance (1.1%) from the vapour over solid paraldehyde at 20°C*

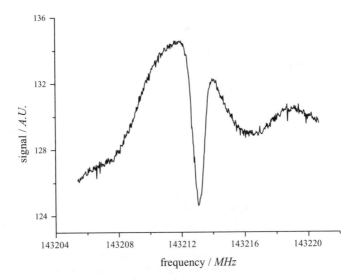

Figure 6.8 *The $2_{02}-1_{01}$ 143213 MHz transition in $HCH^{18}O$ at 26 Pa in natural abundance (0.2%) from the vapour over solid paraldehyde at 20°C*
(Reprinted from Rezgui *et al.*[1] with permission from Elsevier Science)

series of lines being separated by 0.5–1 GHz in the 27–40 GHz band. The three lines of methyl formate in Figure 6.12 show an interesting effect by displaying simultaneously the absorption at the second and the third harmonic of the locked Gunn source. Whilst this may cause problems in interpretation, for identification

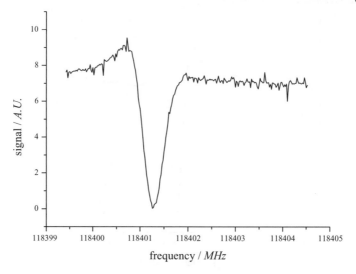

Figure 6.9   *The 4_{14}–3_{13} 118401 MHz transition in ethylene oxide at 13 Pa*

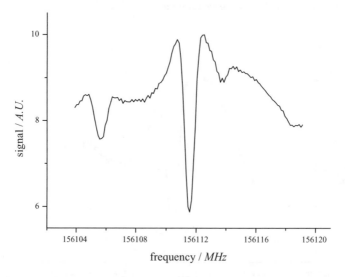

Figure 6.10   *The 8_{18}–7_{17} 156112 MHz transition in acetaldehyde. The line is due to the A-species torsional rotamer at 13 Pa; that due to the other E-species is off the picture. The other lines are due to acetaldehyde*

of molecular species it is particularly advantageous, if somewhat fortuitous! The effect can be readily removed by placing a high-pass MMW filter* between the source and the cavity. The spectrometer's ultimate performance was the observation of a vibrationally excited isotopomer of OCS, itself present in a pure OCS

---

*This comprises a short piece of tapered waveguide such that its cut-off frequency is in between the frequencies of the lower and upper harmonics.

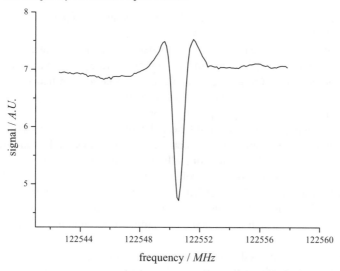

Figure 6.11 *The $9_{36}-8_{27}$ 122551 MHz transition in the trans-rotamer of ethanol at 8 Pa*

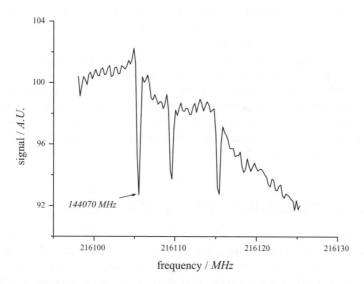

Figure 6.12 *Three transitions of methyl formate at 3 Pa. Displayed simultaneously is the 144070 MHz line due to the $21_{3,18}-21_{2,19}$ transition in the A-torsional species and the 216110 and 216116 MHz lines due to the $19_{2,18}-18_{2,17}$ A- and E-torsional species*

sample at 2.7 ppm abundance without integration, and of an air-diluted sample containing this same species at 360 ppb after computer processing.[2]

A simple gas handling system sufficed to give the performances quoted. The main cell was evacuated by the combination of a two-stage vacuum pump backing a turbomolecular pump to a pressure below 1 Pa and samples flowed slowly through at $\sim$0.01–0.5 L min^{-1} and pressure $\sim$10 Pa.

Flow rate could be increased in order to decrease response time for rapidly fluctuating concentration, by increasing the pumping rate. The extra cost of the larger turbomolecular pump needed to achieve this was not justified in our work. For pressures above $\sim$13 Pa it was possible to achieve reasonable flow rates into the cavity using the two stage pump only. If one accepted this sort of limit to the minimum operating pressure, it would be a low-cost practical alternative, with little compromise to the analytical spectroscopy. Gas flow and blending were initially controlled manually or electronically with adjustable needle valves, but in the later stages by mass-flow controllers calibrated for each gas. Pressure measurements were originally recorded with a Pirani gauge, but in subsequent work it was replaced by a Baratron. No attempt was made to regulate the sample or cell temperatures, which remained near ambient throughout all the experiments.

Sampling of the simulated process gas streams in our work was usually by pumping against a needle valve or capillary restriction. Others have pumped against permeable membranes[7] and that would be useful for many applications. Membranes tend to be selective to particular species and calibration becomes proportionately more complex, however. Other useful total sample throughput restrictors for sampling interfaces include the adjustable reed valve, often found in high-speed two-stroke petrol engines, and the poppet valve. This last is the basis for Diesel engine fuel injectors and is useful for sampling gas or liquid streams particularly at high pressure under mechanical or electrical actuator control. Diesel automobile engine fuel injectors are in fact a useful low-cost option for this purpose. More adaptable piezoelectric actuated valves are available with their control systems, capable of repetitively injecting down to sub-μL aliquots of sample into air streams or even directly into the cavity for discrete sample introduction (Chapter 7).

4   Sample Modulation and Signal Detection

Of the many possible forms of signal modulation, that of source FM has the advantage of being simple to apply and requiring no additional MMW components or structural modifications to the absorption cavity. The penalty is the considerable attention that must be paid to the interaction between source, cavity and sample to obtain reproducible results that are capable of analysis. For a start, the use of a cavity presupposes that the source frequency be kept in synchronism with the cavity resonance whenever data are taken, and in respect of achieving this source FM possesses a distinct advantage over other modulation methods.

Our mode of operation was that originally demonstrated by Kolbe and Leskovar.[8] In our cavity spectrometer the effect of frequency modulating the source is to transmit a MMW signal through the cavity to the detector whose signal output amplitude fluctuates at the same modulation frequency. This modulation signal disappears, or rather is converted into its second harmonic frequency, when the source frequency is at the cavity resonance. If the detected fundamental frequency signal is then converted to a DC voltage by phase coherent detection, this voltage will change sign as the source tracks through the resonance. In our system it is amplified and applied to the control of a piezoelectric actuator in such a phase as to drive the cavity mirror spacing in the direction that keeps the cavity resonant frequency synchronous with the source.

The servo feedback loop functions by reducing the amplitude of the FM superposed by the cavity response onto the MMW signal to as small a value as possible. This does not mean, however, that the MMW output becomes unmodulated. Rather, the original modulation is replaced by one at twice the modulation frequency, that goes undetected by the servo system's phase-coherent detector. It may, however, be observed and converted into DC by a second phase-coherent detector responding to twice the modulation frequency. In the system so far described, this DC voltage will correspond to the second derivative of the cavity frequency profile (Section 4.1) and will therefore take on a fixed value characteristic of the cavity-Q and its coupling coefficient.

The novel feature of the operation occurs if the source scan-range includes a fixed frequency spectral absorption. This absorption will be enhanced as the cavity resonance is tracked through it, by an increase in the effective path length. At this point the absorption is detected by the second phase-coherent detector that sees it as a variation of the second harmonic of the modulation frequency superposed on the fixed cavity signal. Thus the spectrometer will respond to all spectral features in the range through which the source is tracked rather than to a single one located at whatever frequency the cavity happens to be resonant.

Although these cavity locking, tracking and sample detection processes may sound quite complicated, they occur automatically without any need for operator intervention, and give a sample signal not dissimilar to that produced by second harmonic demodulation of the sample absorption signal in a conventional FM waveguide spectrometer. The only visible difference is the presence of a fixed or sloping cavity background on which the spectral signal is superposed. This replaces the much larger resonant profile that would otherwise be displayed by a fixed dimension cavity subjected to a frequency scan (Figure 6.4).

5 Computer Control, Signal Detection and Amplification

Choice of modulation frequency was governed largely by the limitations of our frequency control and tracking system, explained in Section 3.3 and Figure 3.8. To recall, this consisted of a 500 MHz frequency synthesiser with an output of 100 mW, a minimum step size of 10 Hz, a settling time of 30 ms and a wide range of amplitude- and frequency-modulation facilities. Its output was amplified and harmonically mixed with that from a 10–15 GHz YIG oscillator, and the beat between the two then applied to a 20 MHz IF synchroniser whose DC correction voltage was used to synchronise the two. The resulting stabilised signal was further harmonically mixed with the Gunn source typically higher in frequency by a factor of 5 or 6, that was then stabilised by applying the mixer output to a further synchroniser whose output controlled the Gunn power supply current.

The Gunn oscillator frequency-lock became progressively more unstable as its slewing rate increased, whether caused by rapid frequency stepping or by the application of FM. We found ourselves limited in practice to a stepping time of 50–100 ms and an FM deviation of 10 MHz at a rate of 1 kHz. This was considered reasonably satisfactory in that a complete scan could be completed in a few seconds and that our signal detection rate 2 kHz was close to the frequency at which manufacturers, *e.g.* Millitech,[9] specified their detector performance. The same FM rate was used whether our detector was a Schottky barrier mixer diode or the helium-cooled bolometer.

The main signal detection system was a commercial EG & G phase-coherent detector that consisted of a broadband preamplifier of adjustable 30–60 dB gain that transmitted both 1 kHz and 2 kHz signals from the MMW detector, but was selectively tuned to 2 kHz. The outputs from this were split, with one half input to a home-made 1 kHz phase detector. This generated the piezoelectric actuator control signal that scanned the cavity. The other half went to the EG & G precision low-noise 2 kHz phase sensitive amplifier that passed the spectral to the computer for processing and display (Figure 2.1).

The output from the 1 kHz phase detector was returned through an adjustable low-pass filter to a control unit capable of generating the required 150 V drive for the piezoelectric actuator. The parameters of this filter depended somewhat on the signal to noise ratio of the cavity response signal transmitted by the preamplifier; it requiring a longer time constant and giving a slower response as this deteriorated. It was always readily possible, however, to generate a servo lock by observing on an oscilloscope the disappearance of the 1 kHz signal at the preamplifier output, and its replacement by one at double the frequency as the servo system took control.

The output from the 2 kHz phase detector required much less attention, other than care to ensure that neither the detector input nor the DC output overloaded either instrument. Normally these parameters and the detector phase controls would be set up at the start of a day's work and would need no further adjustment unless gross changes were made to the MMW source frequency.

The support electronics for the spectral source and detector used by the authors has been usually made in-house. Great care has to be exercised when building the circuits, to build in protection for the devices, particularly in the simple matter of switching on- and off bias current supplies to the MMW devices and associated electronics. Bias supplies should always incorporate a low-noise control potentiometer to reduce the bias supply to zero before switching. This minimises the risk of switching transients damaging the detectors or sources. Switches should where possible be *make before break*, with the device being short-circuited before the power supply rail is eventually interrupted; once switched off the MMW semiconductors should be terminated as close as possible to them with a coaxial short-circuit. All leads to the devices should be in coaxial cable with the screen and the waveguide structure at actual ground potential.

The devices are all static electricity sensitive as mentioned in Chapter 2. One should not exaggerate the problem in handling the components but they do require some attention. Over several decades we have lost a few devices to what were either switching- or static-transients. On one occasion whilst actually measuring a signal, the main laboratory fluorescent lighting was switched on, about thirty lamps all firing at once; the detector diode died instantly. If faced with poor, noisy electrical lighting and main supply circuits, a battery operated power supply for the detector and multiplier diode is a good option. They are usually quieter and can be kept in a screened enclosure close to the device.

When using modules with the semiconductor mounted by the manufacturer in the waveguide structure, the problems are minimal. The modules will arrive with the device short-circuited. When soldering the power leads to the module initially, take care to keep the device short-circuited until after it has been finally installed, and then cut the shorting wire with snips when it is all connected. When removing the module from the waveguide structure, it should always be short-circuited before disconnecting and during handling, transport and storage.

When handling the discrete unmounted diodes, one has to be much more careful and our best advice is to seek specialist local knowledge and expertise. Be under no illusion, the diodes are very unforgiving! One morning one of us lost three unmounted devices before tea-break. The inquest never revealed the cause but static electricity was the most likely candidate. Days when the relative humidity is low, particularly in winter, we associate with static electricity problems. Certain clothing is more prone to static-problems than other types. If you find you are getting static electricity shocks from furniture, it is probably not a good day to be fitting new sources and detectors!

6   Post-Detection Signal Enhancement

Weak spectral signals passed from the 2 kHz phase-coherent detector to the computer suffered not only from noise but from a sloping background due to variation with frequency of the cavity-Q and the coupling efficiency. It was found that computer algorithms applied to the data collected were rather effective in dealing with both these problems. A wide range of these is described in Press *et al.*[10] One of the most ubiquitous of such techniques is FFT, the data from which are filtered by a Gaussian profile before being subjected to a reverse transformation. The effect of this is to remove noise components that vary at a more rapid rate than this profile, normally chosen to match the expected linewidth. Though effective for smoothing noisy data, this transformation offered nothing to deal with the sloping background.

We found it quite simple instead to write and deploy a Gaussian convolution program that was applied to every data point in turn and took only a fraction of a second to run with our limited number of 512 data points. Although in its simplest form this is mathematically equivalent to the Fourier transformation described earlier, we modified it so that the convolution was taken with the first- or second derivative of a Gaussian profile of arbitrarily chosen width (Sections 4.2 and 4.3). With this we were able to discriminate against pedestal, linear and quadratic backgrounds as well as to smooth out noise, and thereby to pinpoint spectral peaks present at known spectral absorption frequencies (Figures 4.3 and 4.4).

This post-detector integration technique has proved so effective that we now take repeated scans as quickly as the computer control system will allow, relying on noise averaging by summation of these followed by an appropriate convolution to bring up any spectral features present. The best algorithms from the point of view of chemical analysis are known to be those which search for profiles of a specified centre frequency and linewidth. To use these effectively calls, however, for the systematic generation of a large database of spectral line parameters in addition to their centre frequencies alone, which has not yet *to our knowledge* been carried out for MMW spectra.

7   Quantitative Interpretation of Data

MMW spectra possess a number of features that make direct readout of sample concentration from a measurement less than straightforward. Although the peak absorption coefficient of the sample may be determined in several ways itemised below, Equations 1.24 and 1.48 show that additional information about linewidth, temperature, occupation of vibrational states and even Doppler broadening is required to obtain the analyte concentration. Such calculations can be incorporated into any analytic program, but the parameters required are often not readily available in the literature and may need to be determined directly.

This applies most particularly to the linewidth broadening parameter, which can vary with the sample composition by up to an order of magnitude. The values for concomitant gases on the NH_3 3,3 line range typically from 10 (He) to 205 (NH_3) kHz Pa^{-1} (ref. 11, p. 364). This parameter reflects the different cross-sections experienced by the two molecules in collision. Its effect on the analytical

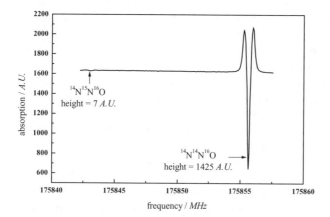

Figure 6.13   *Frequency scan showing the ground state spectral lines of $^{14}N^{15}N^{16}O$ at 175843 MHz and $^{14}N^{14}N^{16}O$ at 175855 MHz, at 1.6 Pa*
(Reprinted from Baker *et al.*[3] with permission from Elsevier Science)

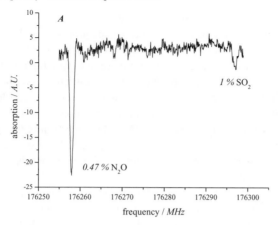

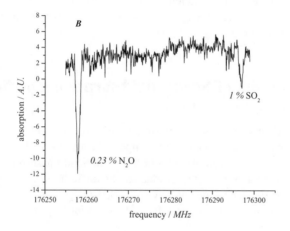

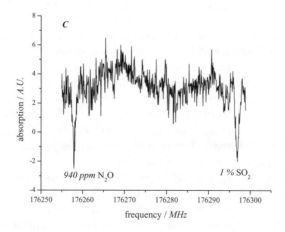

Figure 6.14 *Comparison between signals from N_2O in its 01^10 state and the $41_{4,38} \leftarrow 40_{5,35}$ transition in SO_2 for changing concentrations of N_2O in air*

signal can be much reduced or removed, however, by making a measurement of the line area rather than its peak absorption (Sections 4.1 and 6.8). A rather simple procedure for determining the absorption coefficient of a dilute sample which avoids these complications is to compare its signal with that from an isotopomer or vibrationally excited state in a reference sample of known composition. Most spectral lines are accompanied by satellites due to those sub-species, whose relative concentrations are precisely known. An illustration is in Figure 6.13 in which lines of N_2O and $^{14}N^{15}NO$ present in 3650 ppm natural abundance appear on the same spectral scan. Figure 6.14 shows a comparison carried out between two adjacent spectral lines of the species N_2O and SO_2 at different relative concentrations in air.

Although some correction must still be made for changes in the linewidth broadening parameter, the calculations required are much less detailed. In particular, practical measurements are much simplified by exploiting the insensi-tivity of the peak absorption coefficient to pressure (Section 1.2). This technique becomes especially effective if combined with the line profile integration de-scribed in the next section.

8 Determination of Spectral Absorption Coefficients and Calibration Curves

For reasonably intense spectral lines determination of the absorption coefficient is a fairly straightforward matter. One may compare the cavity MMW transmission without (T_0) and with (T), a sample present at a sufficiently high pressure that its linewidth is much greater than the resonance bandwidth of the cavity.[12] It is necessary then only to determine the cavity loaded-Q, Q_L, to obtain the optical pathlength traversed through the sample from the equations:

$$T/T_0 = \exp(-\gamma_{max} L_{eff}) \approx 1 - \gamma_{max} L_{eff} \text{ and } L_{eff} = Q_L \lambda / 2\pi \qquad (6.1)$$

Both of these are routine tasks when FM is employed. The relative magnitudes of the cavity background signals with- and without the sample present give T/T_0. Q_L is determined from a scan, for which the source and cavity frequencies are maintained at the sample resonance whilst the FM deviation is scanned de-crementally from its upper value to near zero. The servo loop is unable to maintain the lock at zero deviation, hence the direction of scan. Examples of a series of scans for various values of the FM depth on the cavity at resonance frequency 175 GHz are shown in Figure 6.15.[3] These *cavity scans* show a broad maximum whose value reflects the Q of the cavity.

The resulting data have been found to give consistently good fits to a formula derived by Baker *et al.*,[2,3] making it possible to determine the loaded-Q to a precision of 0.1%. The derivation of this formula is not trivial and will not be reproduced here. The maximum of the cavity scan curve described there occurs when the ratio m of the FM deviation Δv_m to the cavity resonance bandwidth equals 2.2. This can be used simply to find an approximate value for the Q. As Q

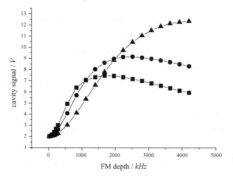

Figure 6.15   *Cavity response vs. modulation depth at $Q = 120\,000$ ■; $Q = 80\,000$ ●; and $Q = 40\,000$ ▲. The system becomes unstable at zero FM depth, so the scan is from high to low frequency, ending ~30 kHz*
(Reprinted from Baker *et al.*[3] with permission from Elsevier Science)

is identical to the ratio of the cavity resonance centre-frequency f_0 to the cavity FWHM, the terms are related through

$$Q_L = mf_0/2\Delta\nu_m \tag{6.2}$$

which at the maximum $[\Delta\nu_{m,max}, V_{max}]$ of the cavity scan becomes

$$Q_L = 1.1f_0/\Delta\nu_{m,max} \tag{6.3}$$

This result may be used in conjunction with Equations 2.4 and 2.5 to determine absolute absorption coefficients. When the sample line is sufficiently weak that the difference between T and T_0 falls below a few percent, rather more care is required. Even after smoothing and computer enhancement, to obtain a significant ratio between these two large and approximately equal quantities is fraught with uncertainty, and some more direct means of observing their difference is necessary. Ideally one would like to amplify the difference between the two without overloading the preamplifier.

One method for achieving this is to adjust the gain of the amplifier so that at one setting it reads the background and at another, higher, setting it reads the signal. With some care in amplifier calibration this can give reliable results. It must be realised, however, that in the case of an FM spectrometer measurement these two signals will not necessarily show the same proportionality to power input and power absorbed, due to the differing linewidths of cavity and of the sample, and that further pre-calibration will be needed.

A widely used alternative method for achieving this is to sweep the spectral source frequency through a range that includes the sample absorption profile, but is sufficiently large for this signal to vanish at the extremities. The data obtained are then fitted to the expected spectral profile plus a background. Although a similar technique has been used successfully by other workers[13,14] we have found the data obtained from an FM spectrometer rather sensitive both to cavity coupling background and the operating conditions.

The former may be dealt with by using computer algorithms together with pre-stored blank runs, but the latter requires careful monitoring of the operating mode of the spectrometer. The critical parameters are the ratios of the FM depth, the cavity width and the linewidth. We have derived expressions for the effects of these on the spectral signal which show that the largest output occurs when all three are comparable[2,3] and is therefore pressure dependent. This dependence can be removed only by working at linewidths much greater than the cavity width, requiring scan ranges larger than those achievable by our tracking system Section 3.3.

It is, however, possible to integrate a spectral signal obtained at lower pressures, over the entire scan range. Figure 6.16 shows a series of scans corresponding to various ratios of FM depth to cavity width.

At low FM depths the spectral peak shows a second derivative form, but as the FM increases the sidelobes characteristic of that form gradually decrease, and at the optimum amplitude almost disappear. In this region the peak becomes exceptionally narrow, making its integration a simple matter. The treatment by Baker *et al.*[3] confirms this behaviour and demonstrates that the area of the line follows the same formula as the peak height of the cavity-generated FM signal. This demodulated signal is itself the cause of the fixed cavity background superposed on the tracked spectral signal. It is thus feasible to obtain a quantity independent of the FM depth, by taking the ratio of the *spectral area* and this fixed *cavity background*. From knowledge of the cavity Q, it is then possible in principle to determine directly the sample partial pressure, without recourse to calibration samples. This result will be proportional to the total sample pressure, so monitoring of the latter is of course called for to obtain a sample concentration.

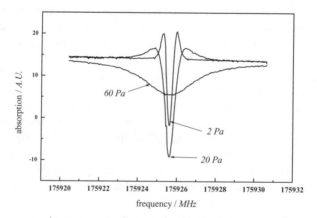

Figure 6.16   *Scans of the spectral line of the $01^{-1}0$ vibrational state N_2O transition at 175925 MHz at increasing pressure and hence linewidth. The FM depth was constant at 264 kHz, and the curves represent therefore decreasing FM depth/linewidth ratio*
(Reprinted from Baker *et al.*[3] with permission from Elsevier Science)

A precise knowledge of the spectral constants and absorption coefficient will also be required to obtain an absolute result.

A trial of these procedures was carried through for samples of oxygen in air and other diluent gases at the 60.306 GHz line.[4] Recalling that, because oxygen possesses a magnetic dipole moment its spectral lines are split into many components by the Earth's magnetic field (Figure 1.2a), in order to carry out quantitative spectrometry the absorption cell was mounted inside a cylindrically folded sheet of the high permeability alloy *Mumetal*. This shields it from the Earth's field so effectively that the many lines collapse into a single one at 60306 MHz (Figure 1.2b). Measurements were made in this magnetically shielded regime on mixtures over the range 0–100% oxygen diluted with nitrogen and carbon dioxide. The pressure dependence response curves showed no significant differences in shape whether calculated from the peak height, total area or area excluding sidelobes. The analytical growth curves for peak height *vs.* oxygen concentration were linear up to ~25% v/v and showed some convex curvature above ~60% oxygen. There was no measurable difference between the shapes of the curves for the nitrogen- and carbon dioxide-diluted samples.

The spectral height response was fitted to two models in an attempt to explain the curvature at higher concentrations. The first was a conventional *Beer–Lambert* relationship of the form:

$$\log T_0/T = \alpha_{max} CL \qquad (6.4)$$

Where C is the oxygen concentration; the other terms have been described previously.

This alone, however, yielded a value for the product $\alpha_{max} L$ that was 3–4 orders of magnitude too large to account for the observed curvature. There was clearly at least one other contributory factor to the curvature and that was found to be due at least in part to collisional broadening effects. The spectral line collisional broadening effect in a two component mixture can be described by

$$1/\Delta\nu_p = x/\Delta\nu_{oxy} + (1-x)/\Delta_{con} \qquad (6.5)$$

where the subscripts refer to p the oxygen apparent linewidth in the mixture; oxy, pure oxygen and, con, the concomitant gas; x is the fractional abundance of oxygen.

The linewidth broadening parameter may be incorporated into the response function derived by Baker *et al.*[3] (their Equations 9 and 12). It relates the area under the spectral line profile *Area*, measured at 2ω, to the spectroscopic parameters and the ratio m of Equation 6.2 above through a term $s = (1 + m^2)^{\frac{1}{2}}$

$$Area = \frac{2x\alpha_{max}QP_{out}c\Delta\nu_s(s-1)}{\pi\nu_0 s(s+1)} \qquad (6.6)$$

Terms have the meanings ascribed previously (Equations 1.48, 4.1).

$\Delta\nu_s$ is the broadened linewidth at the prevailing pressure and would be replaced by $\Delta\nu_p$ which contains the fractional abundance term x. The result is algebraically rather complicated[4] but can be readily programmed into a computer for facile calibration curve plotting. This model gave a more plausible explanation of the growth curve shape with good correlation between the response and the oxygen concentration over the full range 0–100%.

The apparent oxygen absorption linewidth pressure dependence in air derived from the fitting to this simple model showed an effect of greater FM deviation as it became comparable with the narrower spectral line at low pressures. This phenomenon is because of the interaction between the electric field of the modulated wave and that of the relaxing molecule during its transition (ref. 15, p. 139, and ref. 20); see Section 1.2. The effect is to increase the measured spectral absorption linewidth because the molecules are oriented in LTE with the instantaneous electric field due to the source, as they relax after a collision. As the FM deviation increases, so therefore does the bandwidth of the radiation field they experience, and this causes an increase to the linewidth.

The response became linear at higher pressures with a gradient of 18.2 kHz Pa^{-1} rather larger than those reported by Liebe;[13] for the 59.6 GHz line the value he reported was 13.7 kHz Pa^{-1}. Our observations are consistent with area measurements being taken at or near the optimum FM deviation and with a greater linewidth for oxygen self-broadening than concomitant gas broadening, which is the norm. A number of workers, *e.g.* refs. 11, 13, 14, give tables of line broadening *vs.* pressure parameters, although they would be mainly for binary mixtures. For complex gas mixtures the rigorous values will need to be measured. Although one may use an empirical formula of the type exemplified in Equation 6.5, second-order interactions may well create complications, particularly at higher pressures and on the threshold of condensation or chemical interactions.

Considerably more work is required to put the results on a fully quantitative basis. It is quite likely that the simple Equation 6.5 is not up to describing properly the broadening interactions, but the empirical fit to the data within the limits of our experimental error is reassuring. Although it may not be immediately apparent from the combined Equations 6.5 and 6.6, the resulting function is not particularly sensitive to the values of the pure- and the concomitant gas broadening coefficients, particularly when they are close in magnitude. Certainly, as a model equation for plotting response against calibration standards for routine analysis, it is quite adequate.

9 Pressure Scanning

An alternative technique with which we have experimented is that of pressure scanning, for determining both the absorption coefficient α_{max} and sample concentration.[2] In this the spectrometer spectral source and cavity resonance frequencies are held at the peak of the analyte sample absorption. The sample pressure is scanned, either by gradually admitting the sample through a flow valve or by pumping out a cell originally containing the sample. In either case the sample pressure may be atmospheric or some convenient lesser value. Typical spectrometer responses are shown in Figure 6.17a–d for oxygen, formaldehyde, sulfur dioxide and acrylonitrile.

The absorption signal seen varies from a small value at lowest pressures to a peak occurring when spectral line and cavity have comparable widths, before

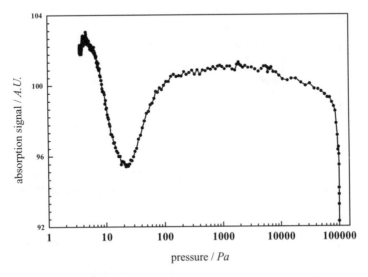

Figure 6.17a   *Pressure scan of oxygen at the J, N = 5,5–4,5 60306 MHz transition*

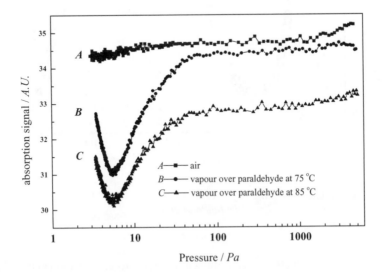

Figure 6.17b   *Pressure scans of formaldehyde at 145603 MHz diluted with air. Paralde-hyde was maintained at elevated temperature under a flow of air at 100 mL min⁻¹ that was sampled into the spectrometer*

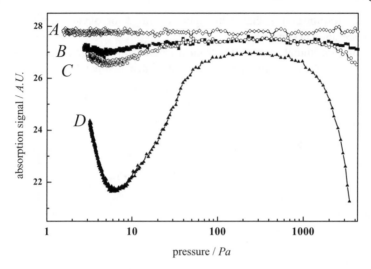

Figure 6.17c   *Pressure scans of SO$_2$/air mixtures at the $51_{8,44}-50_{9,41}$ 145970 MHz SO$_2$ transition for A ◇ air; B ■ 0.12% SO$_2$; C ○ 0.63% SO$_2$; and D ▲ 4% SO$_2$*

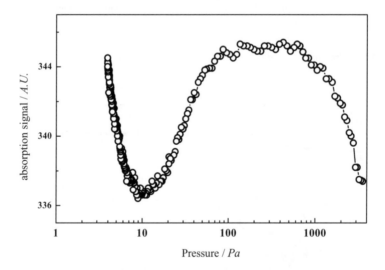

Figure 6.17d   *Pressure scan of acrylonitrile at the $8_{18}-7_{17}$ 154724 MHz transition in nitrogen passed over the liquid at 20°C*

falling off again to a plateau as the linewidth comes to exceed the cavity width. We have been able to derive α_{max} of the sample by comparing the relative heights of peak and plateau, without any need for measuring the sample pressure during the scan. Preliminary results[16] using air and pure oxygen give agreement with the tabulated absorption coefficient[13] to within ~20%. If, however, a concentration determination is required, account must be taken of the changes in spectral linewidth with different gas phase composition and pressure as discussed in Section 6.1.

An especially interesting feature shown by some, but not all of our pressure scans has been that the plateau of absorption observed in our cavity cell between ~150 and 1500 Pa. The plateau absorption starts increasing with further increase in pressure (Figure 6.17a) and in some cases has even produced almost total absorption of the MMW signal. This phenomenon is explained by the increasing overlap with pressure of the absorption signals from wings of further non-resonant spectral lines of the sample. The effect has been well studied and tabulated in the case of water[12] and atmospheric oxygen.[13] Formaldehyde, whose spectrum is sparse, does not show the effect at all in the pressure range studied (Figure 6.17b). Acrylonitrile (Figure 6.17d), whose spectrum is made up of a large number of *b-type* transitions, is so effective an attenuating species that it 'blacks out' the MMW signal before even reaching atmospheric pressure. Each species will show a different threshold pressure at which attenuation will occur. By characterising this threshold pressure at the onset of this increased absorption, the pressure scanning technique may offer a means of distinguishing between species with rich spectra such as acrylonitrile and those with comparatively sparse lines such as water and formaldehyde. As the nett absorption at atmospheric pressure of an individual sample species will also vary with the frequency, it may well be possible also to step a MMW source over a number of longitudinal cavity resonances as has been demonstrated elsewhere.[14] Thus one could obtain a characteristic fingerprint pattern from simple mixtures, from which the identity and even concentration of absorbers might be deduced.

10   The Practical Spectrometer

The goal of creating a MMW spectrometer capable of routine analytical measurement that has lured the authors is now well within reach. The work of ourselves and the many others referenced here and elsewhere has demonstrated the potential of the technique for analysis. Advances in modern computation and electronic engineering coupled with the development of the theoretical basis for the measurements have drawn forward the earlier studies. There is now a sound foundation for the design of compact, automatic MMW spectrometers and the analytical methods to go with them, that are affordable and practically useful to the practising analytical scientist.

References

1. N.D. Rezgui, J. Allen, J.G. Baker and J.F. Alder, Quantitative MMW Spectrometry (I) Design and Implementation of a Tracked MMW Confocal Fabry–Perot Cavity Spectrometer for Gas Analysis, *Anal. Chim. Acta*, 1995, **311**, 99–108.
2. N.D. Rezgui, J.G. Baker and J.F. Alder, Quantitative MMW Spectrometry (II)

Determination of Working Conditions in an Open Fabry–Perot Cavity, *Anal. Chim. Acta,* 1995, **312**, 115–125.

3. J.G. Baker, N.D. Rezgui and J.F. Alder, Quantitative MMW Spectrometry (III) Theory of Spectral Detection and Quantitative Analysis in a MMW Confocal Fabry–Perot Cavity Spectrometer, *Anal. Chim. Acta*, 1996, **319**, 277–290.

4. J.F. Alder and J.G. Baker, Quantitative MMW Spectrometry (IV) Response Curves for Oxygen in Carbon Dioxide and Nitrogen at 60 GHz, *Anal. Chim. Acta,* 1998, **367**, 245–253.

5. N.D. Rezgui, J.G. Baker and J.F. Alder, Quantitative MMW Spectrometry (V) Observation of Dispersive Gas Spectra with a MMW Confocal Fabry–Perot Cavity Spectrometer, *Anal. Chim. Acta,* 2001, **433**, 269–279.

6. G. Thirup, F. Benmakroha, A. Leontakianakos and J.F. Alder, Analytical Microwave Spectrometer Employing a Gunn Diode Locked to the Rotational Absorption Line, *J. Phys. E: Sci. Instrum.*, 1986, **19**, 823–830.

7. R. Varma and L.W. Hrubesh, *Chemical Analysis by Microwave Rotational Spectroscopy*, in *Chemical Analysis*, ed. P.J. Elving, J.D. Winefordner and I.M. Kolthoff, Vol. 52, John Wiley and Sons, 1979.

8. W.F. Kolbe and B. Leskovar, 140 GHz MMW Spectrometer for Chemical Analysis, *Int. J. Infrared & Millimetre Waves*, 1985, **4**, 733–749.

9. Millitech Inc., Millimetre Wave Product Catalogue, 2001, http://www.millitech.com

10. W.H. Press, B.P. Flannery, S.A. Teukolsky and W.T. Vetterling, *Numerical Recipes*, Cambridge University Press, 1986.

11. C.G. Townes and A.L. Schawlow, *Microwave Spectroscopy*, Dover Publications, New York, 1975.

12. Bauer, B. Duterage and M. Godon, Atmospheric Pressure Broadening of the 183 GHz Water Line, *J. Quant Spec. Rad. Trans.*, 1992, **48**, 629–643; *ibid.,* 1985, **33**, 167–175.

13. H.J. Liebe, Updated Model for MMW Propagation in Moist Air: Data on Oxygen and Water Absorption in the Atmosphere, *Radio Sci.* 1985, **20**, 1069–1089.

14. A.F. Krupnov, M.Y. Tretyakov, V.V. Parshin, V.N. Shanin and S.E. Myasnikova, Modern MMW Resonator Spectroscopy of Broad Lines, *J. Mol. Spectrosc.*, 2000, **202**, 107–115.

15. V.L. Vaks, V.V. Kodos and E.V. Spivak, A non-Stationary Microwave Spectrometer, *Rev. Sci. Instrum.*, 1999, **70**, 3447–3453.

16. N.L. Morrey, J.G. Baker and J.F. Alder, unpublished work, UMIST, Manchester, 2000.

CHAPTER 7

The Future for Quantitative Millimetre Wavelength Spectrometry

'To every thing there *is* a season, and a time to every purpose under the heaven.'
Ecclesiastes 3:1

What is then the purpose to quantitative millimetre wavelength spectrometry, and when will be its season?

Analytical methods are created and grow on the need for the measurements they can make. Measurement and analysis are expensive and important, so the methods need to be good, versatile, robust, and in those ways to be better than others. These driving forces dictate at each moment the application of which analytical techniques are applied to the measurement problems of individual scientists and society.

The positive attributes of MMW spectrometry for gas phase analysis include its high selectivity when operated at low pressure. The working pressure can be readily achieved with a modest rotary and turbomolecular pump combination for laboratory and process analysis applications. The pumping requirements will also form an integral part of the essential process of sample acquisition and evacuation.

It is also a technique that has a firm theoretical basis and, as has been demonstrated in this and previous texts, a strong mathematical framework for quantitative analysis. Few methods come so close to being intrinsically absolute in their response to the analyte, with the methods of sensitivity calibration based on quantifiable electrical measurements on the cavity itself.

In the domain of measurements where MMW spectrometry could make an impact, determination of lower molecular weight volatile compounds with well-developed spectra, pollution management stands out as an attainable challenge. Measurement of ultra-trace constituents in the atmosphere is beyond the reach of MMW techniques without pre-concentration. The analytical requirement for pollution management will be driven, however, by legal limits on pollutant emission. In the real world those limits will be low, but not ultra-trace at the point of release, and within therefore the sensitivity of MMW spectrometry. When it comes to the source of pollutant emission: smokestacks, tank-farm vents, engine, rocket and turbine exhausts and the like, there is no shortage of analyte. It is here in the complex, noisy, dirty, wet reality of industrial measurement that cavity MMW spectrometry could well find its niche application.

There is the ability also to achieve quantitative analysis directly on-site using MMW spectrometry with no other calibrant necessary than a supply of fresh air that contains a remarkably constant concentration of oxygen, thus providing the analyst with an abundant reference material.

In field applications of any measurement technique, interference from concomitant species and proper quantification are the major problems, with automation, robustness and minimal maintenance highly desirable features. Here MMW spectrometry can score points over many other techniques. It is certainly amenable to automation and much more robust than the elegant and sensitive optical spectrometric methods that are employed for ultra-trace atmospheric pollutant determinations, with their own problems of fragility, cost and susceptibility to fouling. The scattering problems that beset optical spectrometry so badly are significantly reduced at these longer wavelengths. There is then the possibility of analysing components in aerosol-laden atmospheres with MMW spectrometry.

Others have shown that MMW spectrometry can offer analysis at atmospheric pressure and above for oxygen and water, two very important species in air and indeed in other matrices, particularly simple hydrocarbons. The spectral characteristics of carbon monoxide, nitrogen oxides and sulfur dioxide indicate also their possible analysis at higher pressure. Those species feature strongly in the management of hydrocarbon combustion engines, and MMW spectrometry would be an attractive method for that application.

One more feature is its potential for measuring stereo-, positional and conformational isomers as well as isotopomers in gas-phase species, as was illustrated in Chapter 6. The combination of ion-trap mass spectrometry and ion mobility spectrometry with MMW spectrometry to quantify isomer ratios is surely a proposition worth investigation. The vacuum system in the mass spectrometer is already in place and the ions can be held in position for an extended period in ion traps. There may be time enough to make multiple scans and obtain the spectra of the different isomeric forms or their fragments even at the low abundance characteristic of mass spectrometry systems.

Although the concentration sensitivity of MMW spectrometry is not spectacular compared with other techniques, the mass sensitivity is possibly better than many casual observers would think. Because in a Fabry–Perot cavity the MMW beam is confined to a region near the cavity axis, the most active volume is significantly smaller than the overall volume of the cavity. The authors' calculations based on a cavity $L = 100$ mm working at 150 GHz and 10 Pa are quite revealing. The volume occupied by the electric field of the MMW radiation approximates to a cylindrical prism of radius $\sim(\lambda L/2\pi)^{\frac{1}{2}}$ and volume $\lambda L^2/2$, which at 150 GHz is ~ 10 mL. If the limit of detection for an analyte were, for example, 100 μg L^{-1} at atmospheric pressure, the mass of analyte at the limit of detection in that volume element of the cavity at 10 Pa would be ~100 pg. That would be a quite respectable mass sensitivity in analytical terms, even though the concentration sensitivity were not brilliant. Allowing for imperfection with the result that 10 ng was involved in achieving that concentration limit of detection, even that would be still a useful mass sensitivity.

Whilst admitting the need to actually transfer the gas to that region effectively and to keep it there during the measurement may not be without difficulty, this points the way to a useful technique. It is common practice in many atmospheric gas sampling techniques, and indeed in the analysis of drug and explosives particulates, to cryofocus vapour onto a cold-spot before releasing the concentrate into a chromatograph or spectrometer. There is no reason why this could not be done also into a Fabry–Perot cavity to exploit the high selectivity and quite attractive mass sensitivity for molecules with a MMW spectrum. Maintaining the gas sample introduced into the cavity by flow regime control and minimising lateral diffusion would keep it in the most sensitive region of the cavity field.

It is to be noted that carbon dioxide, one of the most favoured supercritical fluids, is also one of the few gases that are not rotationally active. It is used for the extraction of volatile components in aroma and flavour analysis, a subject area where stereo- and conformational isomerism in the gas phase really is of interest! A supercritical fluid eluate could be passed without problem directly into a MMW cavity spectrometer carrying the aroma or other component with it, to achieve stereo- and conformation specific measurements on smaller (<200 Dalton) molecules.

The tremendous advances in MMW technology over the last few decades were spawned by the development of military radar systems and more importantly wideband telecommunication systems. At one stage in the 1970s it was debatable whether broad bandwidth trunk communication networks would be run in fibre optic or hollow cylindrical waveguide; the rest is history. Trade journals into the eighties still ran this debate[1] and in recent years the advent of *wireless over fibre* as a concept developed along with advances in photoelectronics. The huge anticipated developments in on-line commerce and the wideband digital networks that go with it, along with dozens of channels of TV on optical fibre cable to every home, has created a dynamic research effort into this subject.

Holleitner *et al.*[2] reported microwave spectroscopy on a double quantum dot with an on-chip Josephson oscillator, with potential for use up to 600 GHz. Recent publications tell of 60 GHz full-duplex radio-on-fibre systems[3] and papers are appearing each week talking of advances in this area. There has got to be a tremendous opportunity to piggy-back on a worldwide fibre and wireless communication network. The analytical scientist could send tailored MMW signals to spectrometer nodes all over the planet, or at least across the industrial estate, to monitor that vent stack, or to vehicle testing stations for exhaust gas analysis.

Using a fibre network the encoded MMW spectral source signal could be modulated on to the ~1.6 μm infrared optical carrier along with all the requisite control and identifier signals and sent to the spectrometer node. That would comprise the absorption cavity, photoelectronic demodulator and circuits with low-frequency functions to measure input intensity and time *etc*. Then the demodulated MMW signal would be passed through the cavity. Using *say* atmospheric oxygen as an internal standard to calibrate the instrument sensitivity, the measurement of the target analyte would be made and the measurement encoded and transmitted back through the network to the central computer. The nodes would therefore be relatively low technology hardware with all the MMW and data processing functions at the central location shared amongst the out-stations. No other analytical method offers that capability of intelligent responsive analysis under central computer control for remote monitoring and there is scope for some really ingenious developments.

There is then a *purpose* for MMW spectrometry within the array of modern analytical methods. Through the efforts of the scientists and engineers over 60 years who have provided us with such a wealth of knowledge and diversity of components, all the ingredients are there to develop this powerful technique. Now is *the season* for Quantitative Millimetre Wavelength Spectrometry to grow into the elegant maturity that befits it.

References

1. G.D. O'Clock, C. Hendrickson and W. Schaeffer, Optical and MMW Links Face-Off for the Future, *Microwaves*, 1982, **21**, 59–63.
2. A.W. Holleitner, H. Qin, F. Simmel, B. Irmer, R.H. Blick, J.P. Kotthaus, A.V. Ustinov and K. Eberl, On Chip MMW Spectroscopy of Quantum Dots, *New J. Phys.*, 2000, **2**, 2.1–2.7.
3. T. Kuri, K. Kitayama and Y. Takahashi, GHz-Band Full Duplex Radio on Fibre System, *IEEE Photonics Technol. Lett.*, 2000, **12**, 419–421.

Subject Index

Absorption coefficient, 5 *et seq.*, 19, 53, 76, 90, 104–107, 112
Algorithms, filtering, 70
Ammonia, inversion spectrum, 18, 19, 46, 81
Amplitude modulation, 56, 63, 82, 86
Analyte, partial pressure, 108
Atmospheric pressure spectrometry, 72, 76, 85, 113, 116

Background correction, 35, 60, 66–79, 84–85, 101–110
Backward wave oscillator (BWO), 38, 44, 80, 83, 85, 90
Band spectra, rotational, 18
Beam waist radius, 24, 92, 94
Beat frequency, 49, 62, 101
Bias current, switching, 102
Bolometer, He cooled, 57 *et seq.*, 83, 103
Boltzman equation, exponent value, 3
Boltzmann occupancy factor, 9
Broadening
  cell wall collision, 14
  collisional, 11, 14, 109
  Doppler, 7, 10, 11, 66, 68, 104
  Lorentz, 7, 9, 12, 13, 66, 72
  power saturation, 3, 14, 16
  pressure, 12, 65, 68
BWO, *see* Backward wave oscillator

Calibration on fresh air, 116
Carcinotron, 44
Cavity
  beam waist radius, 24
  confocal point, 23, 26
  coupling, 27–37, 40, 57, 81, 95, 101, 102, 108
  diffraction loss, 15, 25, 92

Fabry–Perot, 15, 23, 52, 76, 79, 80–91, 116
  longitudinal mode, 27, 86, 94, 113
  loss function, 27
  microphony, 93
  mirror reflectivity, 25, 92
  scanning, 35, 45, 49–55, 66–76, 86, 90, 93, 101, 103, 106–112, 116
  semi-confocal Fabry–Perot, 23, 81, 91
  spatial filter, 27, 93
  stability criteria, 23
  zero spacing limit, 25
Circulator, 3-port, 46
Clausius–Mossotti theory, 86
Collisional broadening, 11, 14, 109
Concentration, direct measurement, 104, 108
Concomitant gas, line broadening in mixtures, 110
Confocal Fabry–Perot cavity, 23
Convolution
  Gaussian, 72, 103
  signal data, 72, 86, 103
Coupler, narrow bandwidth, 95
Coupling iris, 15, 35, 40
Critical coupling, 29, 30

Data smoothing, 60, 71, 103
Depth of modulation, 68, 76
Derivative Gaussian convolution, 69, 72
Detectors, 19, 28, 38–63, 71–72, 81–85, 90, 100–103
  phase sensitive, 49, 58, 82, 102
  sensitivity roll-off, 91
Dielectric spectroscopy, 86
Digital modulation, 52, 66, 68–70, 72, 85
Diluent gas, effect on peak signal, 66, 108
Diode
  Impatt, 38, 41–43

RETURN TO: **CHEMISTRY LIBRARY**

100 Hildebrand Hall • 510-642-3753

LOAN PERIOD 1	2	3
1-MONTH USE		
4	5	6

ALL BOOKS MAY BE RECALLED AFTER 7 DAYS.

Renewals may be requested by phone or, using GLADIS,
type **inv** followed by your patron ID number.

DUE AS STAMPED BELOW.

NON-CIRCULATING UNTIL: MAR 0 2 2004		

FORM NO. DD 10   UNIVERSITY OF CALIFORNIA, BERKELEY
2M 4-03   Berkeley, California 94720–6000